Functional Analysis
for Mathematical Sciences

Functional Analysis
for Mathematical Sciences

Masaki Izumi
Kyoto University, Japan

World Scientific

NEW JERSEY · LONDON · SINGAPORE · BEIJING · SHANGHAI · HONG KONG · TAIPEI · CHENNAI · TOKYO

Published by

World Scientific Publishing Europe Ltd.

57 Shelton Street, Covent Garden, London WC2H 9HE

Head office: 5 Toh Tuck Link, Singapore 596224

USA office: 27 Warren Street, Suite 401-402, Hackensack, NJ 07601

Library of Congress Control Number: 2025042221

British Library Cataloguing-in-Publication Data
A catalogue record for this book is available from the British Library.

SUURIKAGAKU NO TAMENO KANSUUKAISEKIGAKU by Masaki Izumi
Original Japanese language edition published by Saiensu-sha Co., Ltd.
1-3-25, Sendagaya, Shibuya-ku, Tokyo 151-0051, Japan
Copyright © 2021, Saiensu-sha Co., Ltd. All rights reserved.

FUNCTIONAL ANALYSIS FOR MATHEMATICAL SCIENCES

ISBN 978-1-80061-832-9 (hardcover)
ISBN 978-1-80061-833-6 (ebook for institutions)
ISBN 978-1-80061-834-3 (ebook for individuals)

For any available supplementary material, please visit
https://www.worldscientific.com/worldscibooks/10.1142/Q0539#t=suppl

Desk Editors: Soundararajan Raghuraman/Srinidhi Murugan

Typeset by Stallion Press
Email: enquiries@stallionpress.com

Dedicated to the late Professor Huzihiro Araki (1932–2022)

Preface

Preface to the English Edition

This is an English translation of a textbook on functional analysis that I originally wrote in Japanese in 2021 based on my lectures at Kyoto University. I omitted a few sentences that make sense only in Japanese, but aside from those and a few stylistic changes for technical reasons, the content is identical to that of the Japanese edition. Although I am solely responsible for the English translation, I would like to express my sincere gratitude to my wife, Mariko Izumi, for her assistance.

Masaki Izumi

Preface to the Japanese Edition

Functional analysis is, in broad terms, a study of the linear mapping between infinite-dimensional spaces. Historically, while it was developed in mathematical analysis such as in the study of integral equations, it has also been closely related to physics since its early stages, as von Neumann showed the spectral decomposition theorem of unbounded self-adjoint operators in order to mathematically formulate quantum mechanics. Not only is functional analysis now used as a fundamental tool in various fields of mathematics beyond mathematical analysis, but it also offers a basic language for dealing with objects in infinite-dimensional spaces in various fields such as engineering.

This book is based on the lecture notes for the courses "Functional analysis" for third-year students and "Advanced functional analysis" for fourth-year students at the Faculty of Science at Kyoto University. Accordingly, it serves as lecture notes for beginners to study essential topics within a limited time and not as an encyclopedic textbook on functional analysis. (For encyclopedic textbooks suitable for beginners, the reader is referred to Conway (1990).) In terms of writing style, I prioritized conveying the atmosphere of the lectures and minimized the reorganization of the contents for the sake of logical efficiency. Therefore, the organization of this book practically follows the flow of the lectures. Corresponding to the two courses, the content is largely divided into two: The first part consists of the first five chapters, and the second part consists of the next five chapters. The first part, except for Chapter 5, deals with the standard contents of basic functional analysis. On the other hand, Chapter 5 is about locally convex spaces, which is a supplemental topic taught in the class beyond the syllabus, and thus beginners may skip this chapter. (I have added the symbol "#" to those sections that can be skipped during the first reading.) The second part is an introduction to operator theory, dealing with the theory of compact operators and the spectral decomposition theorem of self-adjoint operators. The final chapter, Chapter 10, covers several topics that are often omitted in standard lectures due to time constraints. To maintain balance within the book, the Appendix section includes some basic items that were not proved in the chapters.

The prerequisite knowledge for understanding the contents includes rigorous calculus, linear algebra, general topology (primarily, metric spaces), and the basics of complex analysis. Knowledge of measure theory is necessary to understand L^p spaces, basic examples of Banach spaces, and the spectral decomposition theorem. I assume fundamental knowledge of measure theory, such as the basic convergence theorem, but for topics beyond this scope, I have described them in the Appendix. It is preferable to know Fourier analysis in order to understand certain examples and problems.[1]

Since this book is based on lecture notes, many of the problems are the ones given as homework assignments. As a principle in my lectures, I do not provide direct answers, instead offering plenty of hints to help students solve problems on their own. Many students studying advanced mathematics surely aspire to obtain the ability to solve problems by themselves, so it is important for them to overcome the mindset of feeling assured only when given an answer. Thus, this book does not give answers either, but hints are listed at the end of the book. The reader should aim to work toward the solutions by thinking through the explanations provided in the preceding discussion and the hints.

Reflecting on the fact that many fields are related to functional analysis, we find textbooks written on this subject in various styles. I tried to avoid using the arguments unique to operator algebras, which is my specialty, since the lectures on which this book is based were intended for undergraduates. Yet one's character inevitably shows through, and thus this book may be seen as an introductory textbook for functional analysis with a flavor of operator algebras – which, in the end, reflects the best I can offer.

Lastly, I would like to mention a few of my own memories about mathematics textbooks as I myself write one. Near the end of my second year, I found a Japanese translation of Kolmogorov and Fomin's 1975 book at a used bookstore, and I was absorbed in it. It was different from what I had read earlier, written by Japanese mathematicians. I was attracted by their writing style – clarifying their aims at the beginning of their arguments. It is fair to say that I learned the fundamentals of analysis through that book. Akhiezer and Glazman's 1981 two volumes were also impressive. I read them during the seminar supervised by the late Professor Teruo Ikebe in my fourth year. I learned perturbation theory for operators through them;

---

[1]I have placed those problems recommended for the reader to solve when reading the text as "Problems" and more advanced problems as "Exercises" at the end of each chapter.

then, for the first time, I encountered a profound difficulty in mathematics. (Unfortunately, the only available version of that book at present is an English translation of the first edition of the original, and this does not include perturbation theory.) Looking back, it was an invaluable and exciting experience, which could only be provided by superb books, and perhaps those very books now stand behind this one.

Masaki Izumi

About the Author

 **Masaki Izumi** is a Professor in the Graduate School of Science at Kyoto University. He is a leading mathematician in operator algebras and functional analysis, known for his influential contributions to the classification of subfactors, group actions on operator algebras, and the theory of noncommutative Poisson boundaries.

He has received several major distinctions, including the Inoue Prize for Science (2015) and the Autumn Prize of the Mathematical Society of Japan (2010). He was an Invited Speaker at the International Congress of Mathematicians in 2010.

Professor Izumi has played significant roles in academic publishing, serving for many years as Editor-in-Chief of the *Kyoto Journal of Mathematics*. He has also served on the editorial boards of leading journals, such as the *Journal of the Mathematical Society of Japan*, the *Journal of Operator Theory*, and the *Tohoku Mathematical Journal*.

Contents

Notation

The notation and symbols used in this book are explained as follows.

- The symbols $\mathbb{N}$, $\mathbb{Z}$, $\mathbb{Q}$, $\mathbb{R}$, and $\mathbb{C}$ mean the set of natural numbers, the set of integers, the set of rational numbers, the set of real numbers, and the set of complex numbers, respectively. We do not include 0 in $\mathbb{N}$ and write $\mathbb{N}_0 = \{0\} \cup \mathbb{N}$. We set $\mathbb{T} = \mathbb{R}/2\pi\mathbb{Z}$ and $\mathbb{D} = \{z \in \mathbb{C}; \ |z| < 1\}$.
- We mainly use symbols x and y for elements in normed spaces. To avoid confusion, we denote by t the coordinate function of $\mathbb{R}$.
- We always assume that linear spaces are over $\mathbb{C}$ unless otherwise stated. A subspace of a linear space means a linear subspace. We denote by span S the subspace generated by a subset S of a linear space X. Linear independence is defined only for finite sets in many textbooks of linear algebra. In this book, we define that a subset S of a linear space is linearly independent if every finite subset of S is linearly independent.
- For a metric space (X, d), we denote by $B(x, r)$ the open ball $\{y \in X; \ d(x, y) < r\}$ centered at $x \in X$ of radius $r > 0$ and by $\overline{B(x, r)}$ the closed ball. We define the distance between a closed set F and $x \in X$ by $d(x, F) = \inf\{d(x, y); \ y \in F\}$.
- When Ω is a compact Hausdorff space, we denote by $C(\Omega)$ the set of complex-valued continuous functions on Ω. When Ω is a locally compact Hausdorff space, we denote by $C_c(\Omega)$ the set of continuous functions on Ω with compact support. Here, the support supp f of a function f is defined by the closure of $\{\omega \in \Omega; \ f(\omega) \neq 0\}$. When U is an open subset of $\mathbb{R}^n$, we write $C_c^m(U) = C_c(U) \cap C^m(U)$.

By defining $f \in C_c^m(U)$ to be 0 on the complement of U, we may regard f as a function defined on $\mathbb{R}^n$.

- Regarding the notation for a measure space $(\Omega, \mathcal{F}, \mu)$, we often omit the σ-algebra $\mathcal{F}$ and simply denote it by (Ω, μ). When $P(\omega)$ is a condition on $\omega \in \Omega$ and the set of $\omega \in \Omega$ satisfying $P(\omega)$ belongs to $\mathcal{F}$, we simply denote it by $\{P(\omega)\}$ and use the shorthand notation $\mu(\{\omega \in \Omega; \ P(\omega)\}) = \mu\{P(\omega)\}$.

- When Y is a subset of a set X, we denote the indicator function of Y by χ_Y; that is,

$$\chi_Y(x) = \begin{cases} 1, & x \in Y, \\ 0, & x \in X \setminus Y. \end{cases}$$

- When $f : X \to Y$ is a map and X_0 is a subset of X, we denote the restriction of f to X_0 by $f|_{X_0}$.
- When F is a finite subset of a set X, we write $F \Subset X$.

A_k: integral operator with a kernel k.

B_X: the closed unit ball of a normed space X.

$\mathbf{B}(X, Y)$: the set of bounded linear operators from X to Y.

$\mathbf{B}(X) = \mathbf{B}(X, X)$.

$\mathbf{B}(\mathcal{H})_{\mathrm{sa}}$: the set of bounded self-adjoint operators on a Hilbert space $\mathcal{H}$.

$\mathbf{B}(\mathcal{H})_+$: the set of bounded positive operators on $\mathcal{H}$.

$\mathcal{B}(\Omega)$: the set of Borel functions on a topological space Ω.

$\mathcal{B}^b(\Omega)$: the set of bounded Borel functions on Ω.

$\mathfrak{B}_\Omega$: the σ-algebra of Borel subsets of Ω.

$\mathfrak{F}_\Omega$: the set of finite subsets of Ω.

$\mathcal{D}(T)$: the domain of an operator T.

$\mathbf{F}(X, Y)$: the set of finite rank operators from X to Y.

$\mathbf{F}(X) = \mathbf{F}(X, X)$.

$\mathrm{FR}(X)$: the set of Fredholm operators on X.

$\mathrm{FR}_n(X)$: the set of Fredholm operators of index n.

$\mathcal{G}(T)$: the graph of an operator T.

$\mathfrak{K}_\Omega$: the set of compact subsets of Ω.

$\mathbf{K}(X, Y)$: the set of compact operators from X to Y.

$\mathbf{K}(X) = \mathbf{K}(X, X)$.

$\mathbf{K}(\mathcal{H})_{\mathrm{sa}} = \mathbf{K}(\mathcal{H}) \cap \mathbf{B}(\mathcal{H})_{\mathrm{sa}}$.

$\mathbf{K}(\mathcal{H})_+ = \mathbf{K}(\mathcal{H}) \cap \mathbf{B}(\mathcal{H})_+$.

M_f: the multiplication operator of a function f.

$\mathfrak{N}(x)$: the set of neighborhoods of x.

$\mathfrak{O}_\Omega$: the set of open subsets of Ω.

$\mathcal{P}(\mathcal{H})$: the set of projection operators on $\mathcal{H}$.

$\mathcal{R}(T)$: the range of an operator T.

$\mathbf{S}_1(\mathcal{H})$: the trace class operators on $\mathcal{H}$.

$\mathbf{S}_2(\mathcal{H})$: the Hilbert–Schmidt class operators on $\mathcal{H}$.

S_X: the unit sphere of X.

T^*: the adjoint operator of T.

T_f: the Toeplitz operator with symbol f.

$\mathcal{U}(\mathcal{H}_1, \mathcal{H}_2)$: the set of unitary operators from $\mathcal{H}_1$ to $\mathcal{H}_2$.

$\mathcal{U}(\mathcal{H}) = \mathcal{U}(\mathcal{H}, \mathcal{H})$.

V_S: the Cayley transform of a symmetric operator S.

Chapter 1

Basics of Banach Spaces

In this chapter, we begin by introducing Banach spaces, based on which functional analysis is developed, along with many examples. The basic properties of Banach spaces and operators acting between them are also derived.

1.1 Definition and Basic Examples of Banach Spaces

Definition 1.1. A function $X \ni x \mapsto \|x\| \in [0, \infty)$ defined on a linear space X is said to be a *norm* if the following hold for all $x, y \in X$ and $\alpha \in \mathbf{C}$:

(1) $\|x\| = 0$ if and only if $x = 0$.   (positivity)
(2) $\|x + y\| \leq \|x\| + \|y\|$.   (triangle inequality)
(3) $\|\alpha x\| = |\alpha| \|x\|$.   (homogeneity)

The pair $(X, \|\cdot\|)$ is said to be a *normed space*. We often simply say that X is a normed space if the norm is understood. When we would like to explicitly show that the norm is defined on X, we write $\|\cdot\|_X$.

For a normed space $(X, \|\cdot\|)$, if we define $d(x, y) = \|x - y\|$ for $x, y \in X$, the function $d : X \times X \to [0, \infty)$ satisfies the axioms of a metric. In what follows, we regard X as a metric space with this metric unless otherwise stated. The triangle inequality $\|x\| = \|(x - y) + y\| \leq \|x - y\| + \|y\|$, together with the inequality obtained by exchanging the roles of x and y, shows that $\|\|x\| - \|y\|\| \leq \|x - y\|$, and the norm is continuous.

1

Example 1.1. Let $1 \leq p \leq \infty$, and define a norm of $x = (x_1, x_2, \ldots, x_n) \in \mathbb{C}^n$ by

$$\|x\|_p = \begin{cases} (\sum_{i=1}^{n} |x_i|^p)^{\frac{1}{p}}, & 1 \leq p < \infty, \\ \max_{1 \leq i \leq n} |x_i|, & p = \infty. \end{cases}$$

Then, $(\mathbb{C}^n, \|\cdot\|_p)$ is a normed space. For $1 \leq p \leq \infty$, the triangle inequality is nothing but the Minkowski inequality (see Theorem A.1).

Definition 1.2. If a normed space $(X, \|\cdot\|)$ is complete with respect to the metric $d(x, y) = \|x - y\|$, we say that $(X, \|\cdot\|)$ is a *Banach space*.

Example 1.2 (The space of continuous functions). If Ω is a compact Hausdorff space, the set of complex-valued continuous functions $C(\Omega)$ is a linear space by pointwise operations. If we define, for $f \in C(\Omega)$,

$$\|f\|_\infty = \sup_{\omega \in \Omega} |f(\omega)| = \max_{\omega \in \Omega} |f(\omega)|,$$

then $(C(\Omega), \|\cdot\|_\infty)$ is a Banach space.

Proof. Leaving the proof that $(C(\Omega), \|\cdot\|_\infty)$ is a normed space for the reader, we only show the completeness. Let $\{f_n\}_{n=1}^{\infty}$ be a Cauchy sequence in $C(\Omega)$. For fixed $\omega \in \Omega$, we have $|f_n(\omega) - f_m(\omega)| \leq \|f_n - f_m\|_\infty$, and $\{f_n(\omega)\}_{n=1}^{\infty}$ is a Cauchy sequence in $\mathbb{C}$. Thanks to the completeness of $\mathbb{C}$, the limit $\lim_{n \to \infty} f_n(\omega)$ exists, which we denote by $f(\omega)$. To prove the completeness, all we have to show is that f belongs to $C(\Omega)$ and that $\{f_n\}_{n=1}^{\infty}$ converges to f with respect to $\|\cdot\|_\infty$. Thus, it suffices to show that $\{f_n\}_{n=1}^{\infty}$ converges to f uniformly on Ω.

Since $\{f_n\}_{n=1}^{\infty}$ is a Cauchy sequence with respect to $\|\cdot\|_\infty$, we have

$$\forall \varepsilon > 0, \ \exists N \in \mathbb{N}, \ \forall m > \forall n \geq N, \ \forall \omega \in \Omega, \ |f_m(\omega) - f_n(\omega)| < \varepsilon.$$

Letting $m \to \infty$ here, we get $|f(\omega) - f_n(\omega)| \leq \varepsilon$ for the same ε, N, n, and ω as above. This means that $\{f_n\}_{n=1}^{\infty}$ converges to f uniformly on Ω. $\qquad\square$

In what follows, we consider only $\|\cdot\|_\infty$ as a norm of $C(\Omega)$ unless otherwise stated.

Example 1.3 (Sequence spaces). The set of complex-valued sequences is a linear space with $(x_k)_{k=1}^\infty + (y_k)_{k=1}^\infty = (x_k + y_k)_{k=1}^\infty$ and $\alpha(x_k)_{k=1}^\infty = (\alpha x_k)_{k=1}^\infty$. For $1 \leq p \leq \infty$, we define

$$\|(x_k)_{k=1}^\infty\|_p = \begin{cases} (\sum_{k=1}^\infty |x_k|^p)^{\frac{1}{p}}, & 1 \leq p < \infty, \\ \sup_{k \in \mathbb{N}} |x_k|, & p = \infty. \end{cases}$$

Let ℓ^p be the set of sequences $(x_k)_{k=1}^\infty$ with finite $\|(x_k)_{k=1}^\infty\|_p$. Then, $(\ell^p, \|\cdot\|_p)$ is a Banach space.

Proof. We show only completeness in the case of $1 \leq p < \infty$ (we leave the case of $p = \infty$ as an exercise). Let $\{x^{(n)}\}_{n=1}^\infty$ be a Cauchy sequence with respect to $\|\cdot\|_p$ in ℓ^p, and let $x^{(n)} = (x_k^{(n)})_{k=1}^\infty$. For each $k \in \mathbb{N}$, we have $|x_k^{(m)} - x_k^n| \leq \|x^{(m)} - x^{(n)}\|_p$, and $\{x_k^{(n)}\}_{n=1}^\infty$ is a Cauchy sequence in $\mathbb{C}$. The completeness of $\mathbb{C}$ implies that the limit $\lim_{n \to \infty} x_k^{(n)}$ exists, which we denote by x_k. All we have to show is that $x = (x_k)_{k=1}^\infty$ belongs to ℓ^p and that $\{x^{(n)}\}_{n=1}^\infty$ converges to x with respect to $\|\cdot\|_p$. Since $\{x^{(n)}\}_{n=1}^\infty$ is a Cauchy sequence with respect to $\|\cdot\|_p$, we have

$$\forall \varepsilon > 0, \ \exists N \in \mathbb{N}, \ \forall m > \forall n \geq N, \ \|x^{(m)} - x^{(n)}\|_p < \varepsilon.$$

In particular, for a fixed $M \in \mathbb{N}$, we have $\left(\sum_{k=1}^M |x_k^{(m)} - x_k^{(n)}|^p\right)^{1/p} < \varepsilon$.

Letting $m \to \infty$, we get $\left(\sum_{k=1}^M |x_k - x_k^{(n)}|^p\right)^{1/p} \leq \varepsilon$ for ε, N, and n as above. Since M is arbitrary, we have $\|x - x^{(n)}\|_p \leq \varepsilon$. This shows that $x = (x - x^{(n)}) + x^{(n)} \in \ell^p$ and $\lim_{n \to \infty} \|x - x^{(n)}\|_p = 0$. $\qquad \square$

Problem 1.1. For ℓ^∞, show the following:

(1) $(\ell^\infty, \|\cdot\|_\infty)$ is complete.
(2) Let $c_0 = \{(x_k)_{k=1}^\infty \in \ell^\infty; \lim_{k \to \infty} x_k = 0\}$. Then, c_0 is a closed subspace of ℓ^∞. In particular, $(c_0, \|\cdot\|_\infty)$ is a Banach space.

Example 1.4 (L^p spaces). Let (Ω, μ) be a measure space. We introduce an equivalence relation $\sim$ into the set of measurable functions on Ω by defining that two measurable functions f and g on Ω

are equivalent if

$$\mu\{f(\omega) \neq g(\omega)\} = 0.$$

For $1 \leq p \leq \infty$, we set

$$\|f\|_p = \begin{cases} \left(\int_\Omega |f(\omega)|^p d\mu(\omega)\right)^{\frac{1}{p}}, & 1 \leq p < \infty, \\ \inf\{r \geq 0; \ \mu\{|f(\omega)| > r\} = 0\}, & p = \infty. \end{cases}$$

Note that $f \sim g$ if and only if $\|f - g\|_p = 0$. Let $L^p(\Omega, \mu)$ be the set of functions f with finite $\|f\|_p$ modulo the above equivalence relation. Then, $(L^p(\Omega, \mu), \|\cdot\|_p)$ is a Banach space (see Theorem A.2). It is customary that f and its equivalence class are not distinguished in notation.

We call $\|f\|_\infty$ the *essential supremum* of $|f|$. When Ω is a compact Hausdorff space, this symbol denotes the same norm as that used for $C(\Omega)$, which might be confusing. However, since we can easily distinguish the two norms from the context, and they actually coincide for sufficiently nice μ, there is no serious problem usually. We summarize a few remarks about notation for L^p spaces:

- $L^p(\Omega, \mu)$ is sometimes denoted by $L^p(\mu)$ depending on the literature, and p could appear as a subscript.
- When Ω is a measurable subset of $\mathbb{R}^n$ and μ is the Lebesgue measure, we often denote $L^p(\Omega, \mu)$ by $L^p(\Omega)$.
- When μ is a counting measure (i.e., every point has a measure of 1), we denote $L^p(\Omega, \mu)$ by $\ell^p(\Omega)$. We have $\ell^p = \ell^p(\mathbb{N})$.

1.2   Bounded Operators

Throughout this section, X, Y, and Z are normed spaces.

A linear map $T : X \to Y$ is called an *operator*. Although operators could be antilinear or nonlinear maps in some literature, we treat only linear operators in this book. We denote by $\mathcal{R}(T)$ the range (or image) of T. The identity map $x \mapsto x$ of X is denoted by I_X or I. A linear map from X to $\mathbb{C}$ is called a *linear functional*.

We denote by B_X the unit ball $\{x \in X; \|x\| \leq 1\}$ of X and by S_X the unit sphere $\{x \in X; \|x\| = 1\}$ of X. We say that a subset C of X is *bounded* if the following holds: $\exists r > 0, \forall x \in C, \|x\| \leq r$.

Definition 1.3. An operator $T : X \to Y$ is *bounded* if T maps B_X to a bounded subset of Y. This condition is equivalent to

$$\exists M > 0, \ \forall x \in X, \ \|Tx\| \leq M\|x\|.$$

We denote by $\mathbf{B}(X, Y)$ the set of bounded operators from X to Y, and we denote $\mathbf{B}(X, X)$ by $\mathbf{B}(X)$. We call $\mathbf{B}(X, \mathbb{C})$ the *dual space* of X and denote it by X^*.

For $T \in \mathbf{B}(X, Y)$, we define its *operator norm* by

$$\|T\| = \sup_{x \in B_X} \|Tx\|.$$

By definition, we have $\|Tx\| \leq \|T\|\|x\|$ for all $x \in X$.

The reader can see from the following theorem how fundamental the notion of boundedness for operators is.

Theorem 1.1. *For an operator $T : X \to Y$, the following conditions are equivalent:*

(1) *T is continuous.*
(2) *T is continuous at 0.*
(3) *T is bounded.*

Proof. (1) $\Longrightarrow$ (2) is trivial.

(2) $\Longrightarrow$ (3). Since T is continuous at 0, there exists $\delta > 0$ such that $\|x\| \leq \delta$ implies $\|Tx\| \leq 1$. If $x \in X \setminus \{0\}$, we have $\|\delta\|x\|^{-1}x\| = \delta$ and $\|T(\delta\|x\|^{-1}x)\| \leq 1$. Thus, $\|Tx\| \leq \delta^{-1}\|x\|$ holds for all $x \in X$.

(3) $\Longrightarrow$ (1). Since $\|Tx - Ty\| = \|T(x - y)\| \leq \|T\|\|x - y\|$, if $\{x_n\}_{n=1}^{\infty}$ is a sequence in X converging to $x \in X$, the sequence $\{Tx_n\}_{n=1}^{\infty}$ converges to Tx. Thus, T is continuous. $\square$

Next, let us see the structure of the space $\mathbf{B}(X, Y)$. It is a linear space with $(S+T)x = Sx+Tx$ and $(\alpha S)x = \alpha Sx$, and it is a normed space with the operator norm. For example, the triangle inequality can be verified from

$$\|(S+T)x\| \leq \|Sx\| + \|Tx\| \leq \|S\|\|x\| + \|T\|\|x\| = (\|S\| + \|T\|)\|x\|.$$

If $T \in \mathbf{B}(X, Y)$ and $S \in \mathbf{B}(Y, Z)$, we have

$$\|STx\| \leq \|S\|\|Tx\| \leq \|S\|\|T\|\|x\|,$$

which shows that $ST \in \mathbf{B}(X, Y)$ and the submultiplicativity $\|ST\| \leq \|S\|\|Y\|$ of the operator norm.

Theorem 1.2. *Let X be a normed space, and let Y be a Banach space. Then, $\mathbf{B}(X, Y)$ is a Banach space with the operator norm. In particular, X^* is a Banach space.*

Proof. We show the completeness of $\mathbf{B}(X, Y)$. Let $\{T_n\}_{n=1}^{\infty}$ be a Cauchy sequence in $\mathbf{B}(X, Y)$. For each $x \in X$, we have

$$\|T_m x - T_n x\| \leq \|T_m - T_n\|\|x\|,$$

and $\{T_n x\}_{n=1}^{\infty}$ is a Cauchy sequence in Y. Since Y is complete, the limit $\lim_{n\to\infty} T_n x$ exists, which we denote by Tx. As each T_n is linear, so is $T : X \to Y$. To prove the theorem, it suffices to show that $T \in \mathbf{B}(X, Y)$ and that $\{T_n\}_{n=1}^{\infty}$ converges to T with respect to the operator norm.

Since $\{T_n\}_{n=1}^{\infty}$ is a Cauchy sequence in $\mathbf{B}(X, Y)$, we have

$$\forall \varepsilon > 0, \ \exists N \in \mathbb{N}, \ \forall m > \forall n \geq N, \ \forall x \in X, \ \|T_m x - T_n x\| < \varepsilon\|x\|.$$

Letting $m \to \infty$ here, we get $\|Tx - T_n x\| \leq \varepsilon\|x\|$ for ε, N, n, and x as above, and $\|T - T_n\| \leq \varepsilon$ holds. Thus,

$$T = (T - T_n) + T_n \in \mathbf{B}(X, Y),$$

and $\lim_{n\to\infty} \|T - T_n\| = 0$ holds. $\qquad\square$

We say that $T \in \mathbf{B}(X, Y)$ is *invertible* if there exists $S \in \mathbf{B}(Y, X)$ satisfying $ST = I_X$ and $YS = I_Y$. This is equivalent to the condition that T is a bijection and T^{-1} is bounded.

We say that $T \in \mathbf{B}(X, Y)$ is an *isometry* if $\|Tx\| = \|x\|$ for all $x \in X$. If an isometry is surjective, it is invertible, and its inverse is an isometry too. If X is a Banach space and $T \in \mathbf{B}(X, Y)$ is an isometry, the range $\mathcal{R}(T)$ is complete, and so it is closed in Y.

We say that X and Y are *isomorphic* if there exists an invertible in $\mathbf{B}(X,Y)$. We say that X and Y are *isometrically isomorphic* if there exists a surjective isometry in $\mathbf{B}(X,Y)$.

Example 1.5 (Shift operators). Let $1 \leq p \leq \infty$.

- We define the *bilateral shift* U on $\ell^p(\mathbb{Z})$ by $(Ux)_n = x_{n-1}$. Then, U is a surjective isometry.
- We define the *unilateral shift* V of ℓ^p by

$$(Vx)_n = \begin{cases} 0, & n = 1, \\ x_{n-1}, & n \geq 2. \end{cases}$$

Then, V is an isometry but not a surjection.

Remark 1.1. Noting that every injective linear transformation of a finite-dimensional space is automatically surjective, we see that the existence of operators, such as the unilateral shift, is a peculiar phenomenon in infinite-dimensional spaces.

Example 1.6 (Multiplication operator). We assume $1 \leq p < \infty$ for simplicity. For $f \in C[0,1]$ and $h \in L^p[0,1]$, we set $M_f h(t) = f(t)h(t)$. Then,

$$\|M_f h\|_p = \left(\int_0^1 |f(t)h(t)|^p dt \right)^{\frac{1}{p}} \leq \left(\|f\|_\infty^p \int_0^1 |h(t)|^p dt \right)^{\frac{1}{p}}$$
$$= \|f\|_\infty \|h\|_p,$$

which shows that $M_f \in \mathbf{B}(L^p[0,1])$ and $\|M_f\| \leq \|f\|_\infty$. In fact, we can see that the equality holds as follows. Choosing $t_0 \in [0,1]$ with $|f(t_0)| = \|f\|_\infty$, we set $I_n = [t_0 - 1/n, t_0 + 1/n] \cap [0,1]$. We set $h_n = \chi_{I_n}/|I_n|^{1/p}$, where $|I_n|$ is the length of I_n. Then, we have $\|h_n\|_p = 1$ and

$$\|M_f h_n\|_p = \left(\frac{1}{|I_n|} \int_{I_n} |f(t)|^p dt \right)^{\frac{1}{p}} \geq \min_{t \in I_n} |f(t)| \to \|f\|_\infty, \quad (n \to \infty).$$

The following theorem gives one of the most fundamental ways to construct bounded operators. The construction of the Fourier transform on $L^2(\mathbb{R}^n)$ is its typical application.

Theorem 1.3. *Let X_0 be a dense subspace of a normed space X, let Y be a Banach space, and let $T_0 \in \mathbf{B}(X_0, Y)$. Then, there exists a unique extension $T \in \mathbf{B}(X, Y)$ of T_0. Moreover, $\|T\| = \|T_0\|$ holds. If T_0 is an isometry, so is T.*

Proof. Since X_0 is dense in X and T is assumed to be continuous, the uniqueness of T follows if it exists.

For each $x \in X$, we choose a sequence $\{x_n\}_{n=1}^{\infty}$ in X_0 converging to x. The inequality $\|T_0 x_m - T_0 x_n\| \le \|T_0\| \|x_m - x_n\|$ shows that $\{T_0 x_n\}_{n=1}^{\infty}$ is a Cauchy sequence in Y. Since Y is complete, the limit $\lim_{n \to \infty} T_0 x_n$ exists. We would like to define Tx by this limit. Indeed, if $\{x_n'\}_{n=1}^{\infty}$ is another sequence in X_0 converging to x, we have

$$
\begin{aligned}
\|T_0 x_n - T_0 x_n'\| &\le \|T_0\| \|x_n - x_n'\| \\
&\le \|T_0\| (\|x_n - x\| + \|x - x_n'\|) \to 0, \quad (n \to \infty),
\end{aligned}
$$

and Tx does not depend on the choice of the sequence approximating x.

Next, we see that T is linear. Let $a, b \in X$, and let $\{a_n\}_{n=1}^{\infty}$ and $\{b_n\}_{n=1}^{\infty}$ be sequences in X_0 converging to a and b, respectively. Then, $\{a_n + b_n\}_{n=1}^{\infty}$ converges to $a + b$, and

$$
T(a + b) = \lim_{n \to \infty} T_0(a_n + b_n) = \lim_{n \to \infty} (T_0 a_n + T_0 b_n) = Ta + Tb.
$$

We can show that $T(\alpha a) = \alpha Ta$ in a similar way, and T is linear. The boundedness of T and $\|T\| = \|T_0\|$ follow from

$$
\|Tx\| = \lim_{n \to \infty} \|T_0 x_n\| \le \|T_0\| \lim_{n \to \infty} \|x_n\| = \|T_0\| \|x\|.
$$

If T_0 is an isometry,

$$
\|Tx\| = \lim_{n \to \infty} \|T_0 x_n\| = \lim_{n \to \infty} \|x_n\| = \|x\|,
$$

and T is an isometry too. $\qquad \square$

1.3   Finite-Dimensional Spaces and Quotient Spaces

Definition 1.4. We say that two norms $\| \cdot \|_\alpha$ and $\| \cdot \|_\beta$ defined on a linear space X are *equivalent* if the following condition holds:

$$\exists c > 0, \ \forall x \in X, \ \frac{1}{c}\|x\|_\alpha \leq \|x\|_\beta \leq c\|x\|_\alpha.$$

This is equivalent to the condition that the identity map from $(X, \| \cdot \|_\alpha)$ to $(X, \| \cdot \|_\beta)$ is an isomorphism.

More generally, we say that two metrics d_α and d_β defined on a set X are equivalent if the following holds:

$$\exists c > 0, \ \forall x, \forall y \in X, \ \frac{1}{c}d_\alpha(x,y) \leq d_\beta(x,y) \leq cd_\alpha(x,y).$$

When d_α and d_β are equivalent metrics on X, the metric space (X, d_α) is complete if and only if (X, d_β) is complete (show it).

Example 1.7. Since

$$\|x\|_p \leq \|x\|_1 \leq n\|x\|_\infty \leq n\|x\|_p$$

holds for $1 \leq p \leq \infty$ and $x \in \mathbb{C}^n$, the norms $\| \cdot \|_p$, $1 \leq p \leq \infty$, on $\mathbb{C}^n$ are mutually equivalent. Since $\| \cdot \|_2$ gives the usual Euclidean metric on $\mathbb{C}^n$, the normed space $(\mathbb{C}^n, \| \cdot \|_2)$ is complete, and $(\mathbb{C}^p, \| \cdot \|_p)$, $1 \leq p \leq \infty$, are mutually isomorphic Banach spaces.

More generally, the following holds.

Theorem 1.4. *Any two norms on a finite-dimensional linear space are mutually equivalent. In particular, every finite-dimensional normed space is a Banach space.*

Proof. Let X be a finite-dimensional linear space, and choose a basis $\{e_i\}_{i=1}^n$ of X. For $x = \sum_{i=1}^n x_i e_i$, we set $\|x\|_1 = \sum_{i=1}^n |x_i|$, and we introduce a topology into X by $\| \cdot \|_1$. To prove the theorem, it suffices to show that an arbitrary norm $\| \cdot \|$ is equivalent to $\| \cdot \|_1$.

The triangle inequality shows that $\|x\| \leq M\|x\|_1$ with $M = \max_{1 \leq n \leq n} \|e_i\|$ and, in particular, $\|\cdot\|$ is continuous. Since $S = \{x \in X; \|x\|_1 = 1\}$ is compact, the minimal value m of $\|x\|$ on S exists. As $0 \notin S$, the positivity of the norm $\|\cdot\|$ implies $m > 0$. For $x \in X \setminus \{0\}$, we have $\|x\|_1^{-1}x \in S$, and $m \leq \|\|x\|_1^{-1}x\|$. Thus, $m\|x\|_1 \leq \|x\|$. As this also holds for $x = 0$, every $x \in X$ satisfies $m\|x\|_1 \leq \|x\| \leq M\|x\|$, and $\| \cdot \|$ and $\| \cdot \|_1$ are equivalent. $\square$

Corollary 1.1. *Let X and Y be normed spaces, and assume that X is finite-dimensional. Then, every operator $T : X \to Y$ is bounded.*

Proof.   We choose a basis $\{e_i\}_{i=1}^n$ of X, and let $\| \cdot \|_1$ and m be as in the proof of the previous theorem. Letting $M_1 = \max_{1 \le i \le n} \|Te_i\|$, we get the following for $x = \sum_{i=1}^n x_i e_i$:

$$\|Tx\| \le \sum_{i=1}^n |x_i| M_1 = M_1 \|x\|_1 \le \frac{M_1}{m} \|x\|.$$

Thus, T is bounded.   □

Let (X, d) be a metric space. Note that a subset Y of X is closed if the restriction of d to Y is complete.

Corollary 1.2. *Every finite-dimensional subspace of a normed space is closed.*

When Y is a closed subspace of a normed space X, for $x \in X$ and its equivalence class $[x] = x + Y \in X/Y$, we set $\|[x]\| = \inf_{y \in Y} \|x + y\|$, and we call it the *quotient norm* of X/Y. Note that $\|[x]\|$ is nothing but the distance $d(x, Y) = \inf_{y \in Y} d(x, Y)$ from x to the closed set Y.

Problem 1.2. Let Y be a closed subspace of a normed space X, and let $Q : X \to X/Y$ be the quotient map.

(1) Show that the quotient norm is a norm of X/Y.
(2) Show that $\|Q\| = 1$ if $Y \ne X$.

Proposition 1.1. *Let X and Y be normed spaces, and assume that Y is finite-dimensional. Then, for a linear map $T : X \to Y$, the following conditions are equivalent:*

(1) $T \in \mathbf{B}(X, Y)$.
(2) $\ker T$ *is closed.*

Proof.   (1) $\Longrightarrow$ (2). Since T is continuous, $\ker T = T^{-1}(0)$ is closed.
      (2) $\Longrightarrow$ (1). Let $Q : X \to X/\ker T$ be the quotient map. Thanks to the fundamental theorem of homomorphisms, there exists a linear map $\tilde{T} : X/\ker T \to \mathcal{R}(Y)$ such that $T = \tilde{T} \circ Q$. On the other hand,

$$\dim X/\ker T = \dim \mathcal{R}(T) < \infty$$

implies that $\tilde{T}$ is bounded. As Q is bounded, so is T.   □

As a special case of the above proposition, we see that for a linear functional $\varphi : X \to \mathbb{C}$,

$$\varphi \in X^* \iff \ker \varphi \text{ is closed.}$$

The following theorem may be considered as the fundamental theorem of homomorphisms in the category of normed spaces.

Theorem 1.5. *Let X and Y be normed spaces, let $T \in \mathbf{B}(X, Y) \setminus \{0\}$, and let $Q : X \to X/\ker T$ be the quotient map. Then, there exists a unique $\tilde{T} \in \mathbf{B}(X/\ker T, Y)$ satisfying $T = \tilde{T} \circ Q$. Moreover, we have $\|\tilde{T}\| = \|T\|$.*

Proof. The existence and uniqueness of a linear map $\tilde{T} : X/\ker T \to Y$ satisfying $T = \tilde{T} \circ Q$ follows from the fundamental theorem of homomorphisms in linear algebra. Let $a \in X/\ker T \setminus \{0\}$. Then,

$$\forall \varepsilon > 0, \ \exists x \in X, \ [x] = a \quad \text{and} \quad \|a\| \le \|x\| \le \|a\| + \varepsilon.$$

Thus,

$$\|\tilde{T}a\| = \|Tx\| \le \|T\|\|x\| \le \|T\|(\|a\| + \varepsilon).$$

As $\varepsilon > 0$ is arbitrary, we get $\|\tilde{T}\| \le \|T\|$. On the other hand,

$$\|T\| = \|\tilde{T} \circ Q\| \le \|\tilde{T}\|\|Q\| = \|\tilde{T}\|$$

shows that $\|\tilde{T}\| = \|T\|$. $\qquad\square$

Theorem 1.6. *Let Y be a closed subspace of a Banach space X. Then, X/Y with the quotient norm is a Banach space.*

Proof. We show that X/Y is complete with respect to the quotient norm. Let $\{a_n\}_{n=1}^{\infty}$ be a Cauchy sequence in X/Y. Since a Cauchy sequence having a convergent subsequence is convergent, we show that $\{a_n\}_{n=1}^{\infty}$ has a convergent subsequence.

Since $\{a_n\}_{n=1}^{\infty}$ is a Cauchy sequence, we can inductively choose a subsequence $\{a_{n_k}\}_{k=1}^{\infty}$ such that $\|a_{n_{k+1}} - a_{n_k}\| < 2^{-k}$ holds for all $k \in \mathbb{N}$. We choose a representative $x_k \in X$ of $a_{n_{k+1}} - a_{n_k}$ satisfying $\|x_k\| < 2^{-k}$, and we choose a representative $y_1 \in X$ of a_{n_1}. Letting

$y_k = y_1 + \sum_{j=1}^{k-1} x_j$, we get $Q(y_k) = a_{n_k}$, where $Q : X \to X/Y$ is the quotient map. For $k < l$, we have

$$\|y_l - y_k\| = \|\sum_{j=k}^{l-1} x_j\| \le \sum_{j=k}^{l-1} \|x_j\| \le \sum_{j=k}^{l-1} \frac{1}{2^j} < \frac{1}{2^{k-1}},$$

and $\{y_k\}_{k=1}^{\infty}$ is a Cauchy sequence. As X is a Banach space, it converges, and we denote by $y \in X$ its limit. Since Q is continuous, we get

$$Q(y) = \lim_{k \to \infty} Q(y_k) = \lim_{k \to \infty} a_{n_k},$$

and $\{a_{n_k}\}_{k=1}^{\infty}$ converges. $\qquad \square$

The compactness of a bounded closed set no longer holds in an infinite-dimensional space.

Lemma 1.1. *Let Y be a closed subspace of a normed space X with $Y \ne X$. Then, for every $\varepsilon > 0$, there exists $x \in X$ satisfying $\|x\| = 1$ and $d(x, Y) \ge 1 - \varepsilon$.*

Proof. We choose $a \in X/Y$, with $\|a\| = 1$, and a representative $x_0 \in X$ of a satisfying $\|x_0\| < 1 + \varepsilon$. Then, $x = \|x_0\|^{-1} x_0$ has the desired property. $\qquad \square$

Theorem 1.7. *For a normed space X, the following two conditions are equivalent:*

(1) *X is finite-dimensional.*
(2) *B_X is compact.*

Proof. We leave the proof of (1) $\Longrightarrow$ (2) to the reader and show only (2) $\Longrightarrow$ (1). For this purpose, we assume that $\dim X = \infty$ and deduce that B_X is not compact. We take a countable linearly independent subset $\{x_k\}_{k=1}^{\infty}$ of X and set $X_n = \mathrm{span}\{x_k\}_{k=1}^{n}$. Then, since $\dim X_n = n$, the subspace X_n is closed in X and $X_{n-1} \subsetneq X_n$. Thus, we can choose $y_n \in X_n$ satisfying $\|y_n\| = 1$ and $d(y_n, X_{n-1}) \ge 1/2$. Note that we have $\|y_m - y_n\| \ge 1/2$ for $m \ne n$. Since $\{y_n\}_{n=1}$ is a sequence in B_X and any subsequence of it cannot be a Cauchy sequence, it has no convergent subsequences. Therefore, B_X is not compact. $\qquad \square$

In rough terms, there are two approaches to compactness in functional analysis:

- Characterizing (pre-)compact subsets. The Ascoli–Arzelá theorem (Theorem 7.1) is a typical example. The use of compact operators also belongs to this approach.
- Weakening the topology to produce more compact sets. The Banach–Alaoglu theorem (Theorem 3.7) is a typical example.

We end this chapter by showing the existence of unbounded linear functionals of infinite-dimensional normed spaces, which also provides evidence that analysis in infinite-dimensional spaces is completely different from that in finite-dimensional spaces. As this is not particularly a useful result, the reader in a hurry may skip ahead.

A maximal linearly independent subset of a linear space is called a *Hamel basis*. If $\{e_i\}_{i \in I}$ is a Hamel basis, every $x \in X$ can be uniquely expressed as $x = \sum_{i \in I} x_i e_i$, where $x_i \in \mathbb{C}$ is 0 except for finitely many $i \in I$. The uniqueness implies that the map $\varphi_{e_i} : X \ni x \mapsto x_i \in \mathbb{C}$ is a linear functional. Note that a Hamel basis is completely different from an orthonormal basis in a Hilbert space or a Schauder basis in Banach space theory.

Problem 1.3. Show by using Zorn's lemma that for every linearly independent subset S in a linear space X, there exists a Hamel basis of X including S.

Proposition 1.2. *Every infinite-dimensional normed space has an unbounded linear functional.*

Proof. Let X be an infinite-dimensional normed space, and choose a countable linearly independent subset $\{a_k\}_{k=1}^{\infty}$ of X, with $\|a_k\| = 1$, for all k. Let

$$
b_k = \begin{cases} a_1, & k = 1, \\ a_1 + \frac{1}{k} a_k, & k \geq 2. \end{cases}
$$

Then, $\{b_k\}_{k=1}^{\infty}$ is linearly independent too. We choose a Hamel basis S of X including $\{b_k\}_{k=1}^{\infty}$, and we define the linear functional

$\varphi_e : X \to \mathbb{C}$ for $e \in S$ as above. Then,

$$\ker \varphi_{b_1} = \mathrm{span}(S \setminus \{b_1\}).$$

As $\lim_{k \to \infty} b_k = b_1$, we get

$$\overline{\ker \varphi_{b_1}} = \mathrm{span}\, S = X.$$

This shows that $\ker \varphi_{b_1}$ is not closed, and hence φ_{b_1} is not bounded. □

Exercises

Exercise 1.1
Show that $C^1[0,1]$ is a Banach space with the norm $\|f\| = \|f\|_\infty + \|f'\|_\infty$.

Exercise 1.2
Let Ω be a compact Hausdorff space, and let $\omega \in \Omega$.

(1) Show $\chi_\omega \in C(\Omega)^*$, where $\chi_\omega : \Omega \to \mathbb{C}$ is defined by $\chi_\omega(f) = f(\omega)$.
(2) Show that $C_0(\Omega, \omega) = \{f \in C(\Omega);\ f(\omega) = 0\}$ is a closed subspace of $C(\Omega)$.

Exercise 1.3
We say that a continuous function f defined on a locally compact Hausdorff space Ω *vanishes at infinity* if the following holds: For every $\varepsilon > 0$, there exists a compact subset K of Ω such that $|f(\omega)| < \varepsilon$ holds for every $\omega \in \Omega \setminus K$. We denote by $C_0(\Omega)$ the set of continuous functions on Ω vanishing at infinity, and we define a norm of $f \in C_0(\Omega)$ by $\|f\|_\infty = \sup_{\omega \in \Omega} |f(\omega)|$. Let $\tilde{\Omega} = \Omega \cup \{\infty\}$ be the one-point compactification of Ω. Show that $C_0(\Omega)$ is isometrically isomorphic to $C(\tilde{\Omega}, \infty)$. In particular, show that $(C_0(\Omega), \|\cdot\|_\infty)$ is a Banach space.

Exercise 1.4
Let (Ω, μ) be a σ-finite measure space, and let $1 \le p \le \infty$. For $f \in L^\infty(\Omega, \mu)$, we define the multiplication operator $M_f \in \mathbf{B}(L^p(\Omega, \mu))$ of f by $M_f h(\omega) = f(\omega) h(\omega)$.

(1) Show $\|M_f\| \le \|f\|_\infty$.
(2) Show that for every c with $\|f\|_\infty > c > 0$, there exists a measurable subset E of $\{|f(\omega)| > c\}$ satisfying $0 < \mu(E) < \infty$.
(3) Show $\|M_f\| = \|f\|_\infty$.

Exercise 1.5

Let $A(\mathbb{D})$ be the set of continuous functions on $\overline{\mathbb{D}}$ that are holomorphic on $\mathbb{D}$.

(1) Show that $A(\mathbb{D})$ is a closed subspace of $C(\overline{\mathbb{D}})$.
(2) Show that the restriction map $A(\mathbb{D}) \to C(\partial\mathbb{D})$, $f \mapsto f_{|\partial\mathbb{D}}$, is an isometry.

Chapter 2

Basics of Hilbert Spaces

In this chapter, we introduce a Hilbert space, which is a direct generalization of Euclidean spaces and the most useful among Banach spaces, and derive their basic properties.

2.1 Basic Definitions and Examples of Hilbert Spaces

For a complex number α, we denote by $\overline{\alpha}$ its complex conjugate.

Definition 2.1. Let X be a linear space. We say that a function $\langle \cdot, \cdot \rangle : X \times X \to \mathbb{C}$ is an *inner product* if the following holds for all $x, y, z \in X$ and $\alpha \in \mathbb{C}$:

(1) $\langle x, x \rangle \geq 0$. Moreover, $\langle x, x \rangle = 0$ if and only if $x = 0$.
(2) $\langle x + y, z \rangle = \langle x, z \rangle + \langle y, z \rangle$, $\langle \alpha x, y \rangle = \alpha \langle x, y \rangle$.
(3) $\langle x, y \rangle = \overline{\langle y, x \rangle}$.

The pair $(X, \langle \cdot, \cdot \rangle)$ is called an *inner product space* or *pre-Hilbert space*. As in the case of normed spaces, we often call X an inner product space for simplicity. When we would like to emphasize that we consider the inner product of X, we sometimes write $\langle \cdot, \cdot \rangle_X$.

For an inner product, the *Cauchy–Schwarz inequality* holds:

$$|\langle x, y \rangle|^2 \leq \langle x, x \rangle \langle y, y \rangle.$$

The equality holds if and only if x and y are linearly dependent. For the proof, the reader is referred to a textbook on linear algebra.

We can define a norm by $\|x\| = \sqrt{\langle x, x \rangle}$. Indeed, the only non-trivial condition among the axioms of a norm is the triangle inequality, which follows from

$$(\|x\| + \|y\|)^2 - \|x + y\|^2 = 2\|x\|\|y\| - 2\operatorname{Re}\langle x, y \rangle$$

and the Cauchy–Schwarz inequality.

In what follows, we regard an inner product space X as a normed space with this norm unless otherwise stated. The inner product is continuous. Indeed, if $\{x_n\}_{n=1}^{\infty}$ and $\{y_n\}_{n=1}^{\infty}$ are sequences in X converging to $x \in X$ and $y \in X$, respectively, then the Cauchy–Schwarz inequality implies

$$\begin{aligned}
|\langle x_n, y_n \rangle - \langle x, y \rangle| &\leq |\langle x_n - x, y_n \rangle| + |\langle x, y_n - y \rangle| \\
&\leq \|x_n - x\|\|y_n\| + \|x\|\|y_n - y\| \\
&\leq \|x_n - x\|\|y_n - y\| + \|x_n - x\|\|y\| + \|x\|\|y_n - y\|,
\end{aligned}$$

which converges to 0.

Definition 2.2. An inner product space $(\mathcal{H}, \langle \cdot, \cdot \rangle)$ is said to be a *Hilbert space* if it is complete with respect to the norm $\|x\| = \sqrt{\langle x, x \rangle}$.

Example 2.1. Let (Ω, μ) be a measure space. The Banach space $L^2(\Omega, \mu)$ is a Hilbert space with an inner product,

$$\langle f, g \rangle = \int_{\Omega} f(\omega)\overline{g(\omega)}d\mu(\omega).$$

As special cases:

- $\mathbb{C}^n$ is a Hilbert space with the standard inner product $\langle x, y \rangle = \sum_{k=1}^{n} x_k \overline{y_k}$.
- The sequence space ℓ^2 is a Hilbert space with $\langle x, y \rangle = \sum_{k=1}^{\infty} x_k \overline{y_k}$.

Let X be a linear space. We say that a function $f : X \times X \to \mathbb{C}$ is a *sesquilinear form* if the following holds: For all $x, y, z \in X$ and $\alpha \in \mathbb{C}$,

- $f(x + y, z) = f(x, z) + f(y, z)$, $f(\alpha x, z) = \alpha f(x, z)$,
- $f(z, x + y) = f(z, x) + f(z, y)$, $f(z, \alpha x) = \overline{\alpha} f(z, x)$.

An inner product is a sesquilinear form. Using

$$\sum_{k=0}^{3} i^{nk} = \begin{cases} 4, & n \equiv 0 \mod 4, \\ 0, & n \not\equiv 0 \mod 4, \end{cases}$$

where i is the imaginary unit and $n \in \mathbb{N}$, we see that the following *polarization identity* holds for every sesquilinear form f:

$$f(x, y) = \frac{1}{4} \sum_{k=0}^{3} i^k f(x + i^k y, x + i^k y), \quad \forall x, y \in X.$$

The point of this equation is that f is determined by $f(x, x)$ although $f(x, y)$ is a function of two variables.

Proposition 2.1. *For any two elements x and y in an inner product space, the following hold:*

(1) *Parallelogram law:*

$$\|x + y\|^2 + \|x - y\|^2 = 2\|x\|^2 + 2\|y\|^2.$$

(2) *Polarization identity:*

$$\langle x, y \rangle = \frac{1}{4} \sum_{k=0}^{3} i^k \|x + i^k y\|^2.$$

We omit the proof as it is a direct computation. (1) is a necessary condition for a norm of a linear space to be given by an inner product. On the other hand, the Jordan–von Neumann theorem (Jordan and von Neumann, 1935) states that if a norm of a Banach space satisfies (1), we can define an inner product by (2).

2.2 Projection Theorem and Its Applications

Definition 2.3. A subset C of a linear space is said to be a *convex set* if $(1 - t)x + ty \in C$ for all $x, y \in C$ and $0 < t < 1$.

Theorem 2.1 (Projection theorem). *Let C be a closed convex subset of a Hilbert space $\mathcal{H}$, and let $x \in \mathcal{H}$. Then, there exists a unique $x_0 \in C$ satisfying $\|x - x_0\| = \inf_{y \in C} \|x - y\|$. We say that x_0 is the projection of x onto C.*

Proof.   Let $\alpha = \inf_{y \in C} \|x - y\|$, and choose a sequence $\{y_n\}_{n=1}^{\infty}$ in C such that $\{\|x - y_n\|\}_{n=1}^{\infty}$ converges to α. The parallelogram law implies

$$\|x - y_m + x - y_n\|^2 + \|(x - y_m) - (x - y_n)\|^2 = 2\|x - y_m\|^2 + 2\|x - y_n\|^2,$$

and

$$\|y_n - y_m\|^2 = 2\|x - y_m\|^2 + 2\|x - y_n\|^2 - 4\|x - \frac{1}{2}(y_m + y_n)\|^2$$

$$\leq 2\|x - y_m\|^2 + 2\|x - y_n\|^2 - 4\alpha^2.$$

This shows that $\{y_n\}_{n=1}^{\infty}$ is a Cauchy sequence. Since $\mathcal{H}$ is complete and C is closed, the sequence $\{y_n\}_{n=1}^{\infty}$ converges in C, and we denote by x_0 its limit. Then, $\|x - x_0\| = \alpha$. If $x_0' \in C$ also satisfies the same condition, the parallelogram law again implies

$$\|x_0 - x_0'\|^2 = 2\|x - x_0\|^2 + 2\|x - x_0'\|^2 - 4\|x - \frac{1}{2}(x_0 + x_0')\|^2 \leq 0,$$

which shows $x_0' = x_0$.   □

Problem 2.1. Give an example of a Banach space, for which the conclusion of the above theorem does not hold.

In the remainder of this section, $\mathcal{H}$ will be a Hilbert space.

Definition 2.4. If $\langle x, y \rangle = 0$ holds for $x, y \in \mathcal{H}$, we say that x and y are *orthogonal* and write $x \perp y$. For a subset $S \subset \mathcal{H}$, we set $S^{\perp} = \{x \in \mathcal{H}; \ \forall y \in S, \ x \perp y\}$, and call it the *orthogonal complement* of S.

Direct consequences of the above definition are in order:

- $x \perp y$ implies $\|x + y\|^2 = \|x\|^2 + \|y\|^2$ (Pythagorean theorem).
- $(S^{\perp})^{\perp} \supset S$ always holds. $S_1 \subset S_2$ implies $S_2^{\perp} \subset S_1^{\perp}$.
- As the inner product is continuous and

$$S^{\perp} = \bigcap_{y \in S} \{x \in \mathcal{H}; \ \langle x, y \rangle = 0\},$$

$S^{\perp}$ is a closed subspace of $\mathcal{H}$.

- The continuity and linearity of the inner product imply

$$S^\perp = (\text{span } S)^\perp = \overline{\text{span } S}^\perp.$$

Theorem 2.2 (Orthogonal decomposition theorem). *Let $\mathcal{K}$ be a closed subspace of $\mathcal{H}$. Then:*

(1) $\mathcal{H} = \mathcal{K} \oplus \mathcal{K}^\perp$.
(2) $(\mathcal{K}^\perp)^\perp = \mathcal{K}$.

Proof. (1) It suffices to show $\mathcal{K} + \mathcal{K}^\perp = \mathcal{H}$, as we can see that $\mathcal{K} \cap \mathcal{K}^\perp = \{0\}$ from the definition of the norm of $\mathcal{H}$. Let $x \in \mathcal{H}$, and let x_0 be the projection of x onto $\mathcal{K}$. We show $x_1 = x - x_0 \in \mathcal{K}^\perp$. For $y \in \mathcal{K}$, we have $x_0 - y \in \mathcal{K}$, and the definition of x_0 implies $\|x - x_0\| \leq \|x - x_0 + y\|$. Thus,

$$\|x_1\|^2 \leq \|x_1 + y\|^2 = \|x_1\|^2 + 2\,\mathrm{Re}\langle y, x_1 \rangle + \|y\|^2,$$

and $2\,\mathrm{Re}\langle y, x_1 \rangle + \|y\|^2 \geq 0$. Letting t be an arbitrary real number and replacing y with $t\langle x_1, y \rangle y$, we get $|\langle x_1, y \rangle|^2 (2t + \|y\|^2 t^2) \geq 0$. Therefore, $\langle y, x_1 \rangle = 0$ and $x_1 \in \mathcal{K}^\perp$.

(2) Since $\mathcal{K} \subset (\mathcal{K}^\perp)^\perp$ trivially holds, we show only $(\mathcal{K}^\perp)^\perp \subset \mathcal{K}$. Let $x \in (\mathcal{K}^\perp)^\perp$. (1) implies that x is decomposed as $x = x_0 + x_1$, with $x_0 \in \mathcal{K}$ and $x_1 \in \mathcal{K}^\perp$. Since $0 = \langle x, x_1 \rangle = \|x_1\|^2$, we get $x = x_0 \in \mathcal{K}$. $\qquad\square$

Noting that $S^\perp = \overline{\text{span } S}^\perp$ holds for $S \subset \mathcal{H}$, we can see that (2) implies the following statement.

Corollary 2.1. *For $S \subset \mathcal{H}$, we have $(S^\perp)^\perp = \overline{\text{span } S}$. In particular, the following conditions are equivalent:*

(1) $S^\perp = \{0\}$.
(2) span S *is dense in* $\mathcal{H}$.

When $\mathcal{K}$ is a closed subspace of $\mathcal{H}$, the projection x_0 of $x \in \mathcal{H}$ onto $\mathcal{K}$ in the sense of the projection theorem coincides with the projection with respect to the direct sum decomposition $\mathcal{H} = \mathcal{K} \oplus \mathcal{K}^\perp$. Thus, the map $P_\mathcal{K}$ associating x_0 to x is a linear map satisfying $P_\mathcal{K}^2 = P_\mathcal{K}$. We call $P_\mathcal{K}$ the *projection operator* from $\mathcal{H}$ onto $\mathcal{K}$, or simply a *projection*. Note that $I - P_\mathcal{K} = P_{\mathcal{K}^\perp}$ holds. Unless $\mathcal{K} = \{0\}$, we have $\|P_\mathcal{K}\| = 1$.

For $y \in \mathcal{H}$, we set $\varphi_y(x) = \langle x, y \rangle$. Then, φ_y is a linear functional of $\mathcal{H}$, and the Cauchy–Schwarz inequality implies that φ_y is bounded.

The following fact that its converse holds as well makes the theory of Hilbert spaces considerably easier.

Theorem 2.3 (Riesz representation theorem). *Let $\mathcal{H}$ be a Hilbert space, and let $\varphi \in \mathcal{H}^*$. Then, there exists a unique $x_\varphi \in \mathcal{H}$ satisfying $\varphi(x) = \langle x, x_\varphi \rangle$ for all $x \in \mathcal{H}$. Moreover, we have $\|\varphi\| = \|x_\varphi\|$.*

Proof. When $\varphi = 0$, we have $x_\varphi = 0$. We assume $\varphi \neq 0$ now. Since φ is continuous, the subspace $\mathcal{K} = \ker \varphi$ is closed in $\mathcal{H}$. The orthogonal decomposition $\mathcal{H} = \mathcal{K} \oplus \mathcal{K}^\perp$ shows

$$\dim \mathcal{K}^\perp = \dim \mathcal{H}/\ker \varphi = \dim \mathbb{C} = 1,$$

and there exists $y \in \mathcal{K}^\perp$ satisfying $\mathcal{K}^\perp = \mathbb{C}y$. Since $\varphi(y) \neq 0$, we may assume $\varphi(y) = 1$ multiplying y by a constant if necessary. Let $x_\varphi = \|y\|^{-2}y$. For $x \in \mathcal{K}$, we have $\varphi(x) = 0 = \langle x, x_\varphi \rangle$, and we also have $\varphi(y) = 1 = \langle y, x_\varphi \rangle$. As $\mathcal{H} = \mathcal{K} + \mathbb{C}y$, we conclude $\varphi(x) = \langle x, x_\varphi \rangle$ for all $x \in \mathcal{H}$. If x'_φ satisfies the same condition, we get $x_\varphi - x'_\varphi \in \mathcal{H}^\perp = \{0\}$, and $x_\varphi = x'_\varphi$.

The Cauchy–Schwarz inequality implies $\|\varphi\| \leq \|x_\varphi\|$. Since $\varphi(\|x_\varphi\|^{-1}x_\varphi) = \|x_\varphi\|$, we get $\|\varphi\| \geq \|x_\varphi\|$. $\qquad\square$

2.3   Complete Orthonormal Systems

We assume that $\mathcal{H}$ is a Hilbert space throughout this section.

Definition 2.5. Let $S \subset \mathcal{H}$.

- We say that S is an *orthonormal system* if every element of S is of norm 1 and any two elements in S are mutually orthogonal. We abbreviate it as ONS in what follows.
- If S is an ONS and $\overline{\operatorname{span} S} = \mathcal{H}$ (this is equivalent to $S^\perp = \{0\}$), we call it a *complete orthonormal system* or *orthonormal basis*. We abbreviate it as CONS in what follows.

Example 2.2. For $n \in \mathbb{N}$, let $\delta_n = (\delta_{n,k})_{k=1}^\infty \in \ell^2$. Then, $\{\delta_n\}_{n=1}^\infty$ is a CONS for ℓ^2. Here, $\delta_{n,k}$ is the Kronecker delta:

$$\delta_{n,k} = \begin{cases} 1, & k = n, \\ 0, & k \neq n. \end{cases}$$

We call $\{\delta_n\}_{n=1}^\infty$ the canonical basis of ℓ^2.

We recall the Gram–Schmidt orthogonalization procedure. For the proof, the reader is referred to a textbook on linear algebra.

Lemma 2.1. *Let $\{x_k\}_{k=1}^{\infty}$ be a linearly independent subset of $\mathcal{H}$. We set $e_1 = \|x_1\|^{-1}x_1$. For $n \geq 2$, we inductively define e_n by*

$$y_n = x_n - \sum_{k=1}^{n-1}\langle x_n, e_k\rangle e_k, \quad e_n = \frac{1}{\|y_n\|}y_n.$$

Then, $\{e_k\}_{k=1}^{\infty}$ is an ONS satisfying $\operatorname{span}\{x_k\}_{k=1}^{n} = \operatorname{span}\{e_k\}_{k=1}^{n}$ for all n. If, moreover, $\operatorname{span}\{x_k\}_{k=1}^{\infty}$ is dense in $\mathcal{H}$, so is $\{e_k\}_{k=1}^{\infty}$, and it is a CONS for $\mathcal{H}$.

Example 2.3. Since the set of polynomial functions is dense in $L^2[a, b]$ for a finite interval $[a, b]$, applying the Gram–Schmidt orthogonalization procedure to $\{x^n\}_{n=0}^{\infty}$, we get a CONS for $L^2[a, b]$. Such a CONS is called an orthogonal polynomial.

Theorem 2.4. *For an infinite-dimensional Hilbert space $\mathcal{H}$, the following conditions are equivalent:*

(1) *$\mathcal{H}$ is separable.*
(2) *There exists a CONS for $\mathcal{H}$ that is a countable set.*

Proof. (1) $\implies$ (2). Let D be a countable dense subset of $\mathcal{H}$. We choose a linearly independent subset $\{x_k\}_{k=1}^{\infty}$ of D with $\operatorname{span}\{x_k\}_{k=1}^{\infty} = \operatorname{span} D$. Applying the Gram–Schmidt orthogonalization procedure to $\{x_k\}_{k=1}^{\infty}$, we can construct a desired CONS, $\{e_n\}_{n=1}^{\infty}$.

(2) $\implies$ (1). The set of linear combinations of a CONS $\{e_n\}_{n=1}^{\infty}$ with coefficients in $\mathbb{Q} + \mathbb{Q}i$ is a countable dense subset of $\mathcal{H}$. $\qquad\square$

We leave the existence of a CONS for a general Hilbert space as a problem.

Problem 2.2. Show that every Hilbert space has a CONS.

We present important properties of a CONS as follows in order.

Lemma 2.2. *Let $\{e_1, e_2, \ldots, e_n\}$ be an ONS in $\mathcal{H}$. For $x \in \mathcal{H}$, we set*

$$P_n x = \sum_{k=1}^{n} \langle x, e_k \rangle e_k.$$

Then, P_n is the projection from $\mathcal{H}$ onto $\mathrm{span}\{e_1, e_2, \ldots, e_n\}$.

Proof. Let $\mathcal{K}_n = \mathrm{span}\{e_1, e_2, \ldots, e_n\}$. Then, $P_n x \in \mathcal{K}_n$. For $1 \le k \le n$, we have

$$\langle x - P_n x, e_k \rangle = \langle x, e_k \rangle - \langle P_n x, e_k \rangle = 0,$$

and $x - P_n x \in \mathcal{K}_n^{\perp}$. The uniqueness of orthogonal decomposition implies that P_n is the projection from $\mathcal{H}$ onto $\mathcal{K}_n$. $\qquad \square$

Corollary 2.2. *Let $\{e_k\}_{k=1}^{\infty}$ be an ONS in $\mathcal{H}$. Then, the following Bessel inequality holds for every $x \in \mathcal{H}$:*

$$\sum_{k=1}^{\infty} |\langle x, e_k \rangle|^2 \le \|x\|^2.$$

Lemma 2.3. *Let $\{e_k\}_{k=1}^{\infty}$ be an ONS in $\mathcal{H}$, and let $\mathcal{K} = \overline{\mathrm{span}\{e_k\}_{k=1}^{\infty}}$. For $x \in \mathcal{H}$, we set $x_n = \sum_{k=1}^{n} \langle x, e_k \rangle e_k$. Then, $\{x_n\}_{n=1}^{\infty}$ converges to $P_{\mathcal{K}} x$ (in particular, the series $\sum_{k=1}^{\infty} \langle x, e_k \rangle e_k$ converges to $P_{\mathcal{K}} x$ regardless of the order of summation). Moreover, the following holds:*

$$\|P_{\mathcal{K}} x\|^2 = \sum_{k=1}^{\infty} |\langle x, e_k \rangle|^2.$$

Proof. For $m > n$, we have

$$\|x_m - x_n\| = \sqrt{\sum_{k=n+1}^{m} |\langle x, e_k \rangle|^2},$$

and the Bessel inequality implies that $\{x_n\}_{n=1}^{\infty}$ is a Cauchy sequence. Since $\mathcal{H}$ is a Hilbert space, the sequence $\{x_n\}_{n=1}^{\infty}$ converges, and we denote its limit by y. Since $x_n \in \mathcal{K}$ for all $n \in \mathcal{K}$, we have $y \in \mathcal{K}$. Since

$$\langle x - y, e_k \rangle = \langle x, e_k \rangle - \lim_{n \to \infty} \langle x_n, e_k \rangle = \langle x, e_k \rangle - \langle x, e_k \rangle = 0,$$

we have $x - y \in \mathcal{K}^\perp$. Thus, the uniqueness of the orthogonal decomposition implies $y = P_{\mathcal{K}} x$. The continuity of the norm implies

$$\|y\|^2 = \lim_{n \to \infty} \|x_n\|^2 = \lim_{n \to \infty} \sum_{k=1}^{n} |\langle x, e_k \rangle|^2 = \sum_{k=1}^{\infty} |\langle x, e_k \rangle|^2. \qquad \square$$

Remark 2.1. The above lemma shows that if S is a countable ONS in $\mathcal{H}$, the expression $\sum_{e \in S} \langle x, e \rangle e$ makes sense. Fixing an enumeration $S = \{e_n\}_{n=1}^{\infty}$, we may define it by $\sum_{n=1}^{\infty} \langle x, e_n \rangle e_n$.

In the same argument as in the proof of the above lemma, we can show the following.

Corollary 2.3. *Let* $\{e_k\}_{k=1}^{\infty}$ *be an ONS in* $\mathcal{H}$, *and let* $a = (a_k)_{k=1}^{\infty} \in \ell^2$. *Then,* $\sum_{k=1}^{\infty} a_k e_k$ *converges, and its norm is* $\|a\|_2$.

Lemma 2.4. *Let S be an ONS in $\mathcal{H}$. For $x \in \mathcal{H}$, we set $S_x = \{e \in S;$ $\langle x, e \rangle \neq 0\}$. Then, S_x is at most a countable set, and $\sum_{e \in S_x} \langle x, e \rangle e$ is the projection of x onto* $\overline{\text{span}\, S}$. *In what follows, we simply denote it by* $\sum_{e \in S} \langle x, e \rangle e$.

Proof. For $x \in \mathcal{H}$ and $n \in \mathbb{N}$, let $S_{x,n} = \{e \in S;\ |\langle x, e \rangle|^2 \geq 1/n\}$. Then, $S_x = \bigcup_{n=1}^{\infty} S_{x,n}$. For a finite subset F of $S_{x,n}$, the Bessel inequality implies

$$\|x\|^2 \geq \sum_{e \in F} |\langle x, e \rangle|^2 \geq \frac{1}{n} \# F,$$

and $\# F \leq n\|x\|^2$. As this is the case for any finite subset F of $S_{x,n}$, we get $\# S_{x,n} \leq n\|x\|^2$. Thus, S_x is a countable set.

Let $y = \sum_{e \in S_x} \langle x, e \rangle e$, which belongs to $\overline{\text{span}\, S} =: \mathcal{K}$. For $e \in S_x$, we get $\langle x - y, e \rangle = 0$, as in the proof of the previous lemma, and for $e \in S \backslash S_x$, we have $\langle y, e \rangle = 0 = \langle x, e \rangle$. Thus, $x - y \in S^\perp = \mathcal{K}^\perp$, and the uniqueness of the orthogonal decomposition shows $y = P_{\mathcal{K}} x$. $\quad \square$

Theorem 2.5. *Let S be an ONS in $\mathcal{H}$. Then, the following conditions are equivalent.*

(1) *S is a CONS.*

(2) *For any $x \in \mathcal{H}$, we have $x = \sum_{e \in S} \langle x, e \rangle e$.*

(3) *For any $x, y \in \mathcal{H}$, we have $\langle x, y \rangle = \sum_{e \in S} \langle x, e \rangle \langle e, y \rangle$.*

(4) *For any $x \in \mathcal{H}$, we have $\|x\|^2 = \sum_{e \in S} |\langle x, e \rangle|^2$.*

Proof. Let $\mathcal{K} = \overline{\operatorname{span} S}$.

  (1) $\implies$ (2). As $\mathcal{K} = \mathcal{H}$, we get $\sum_{e \in S} \langle x, e \rangle e = P_{\mathcal{K}} x = x$.

  (2) $\implies$ (3). As the proof is easy if S_x is a finite set, we prove the implication when $S_x = \{e_k\}_{k=1}^{\infty}$. Since $x = \sum_{k=1}^{\infty} \langle x, e_k \rangle e_k$, we get

$$\langle x, y \rangle = \lim_{n \to \infty} \left\langle \sum_{k=1}^{n} \langle x, e_k \rangle e_k, y \right\rangle = \lim_{n \to \infty} \sum_{k=1}^{n} \langle x, e_k \rangle \langle e_k, y \rangle.$$

  (3) $\implies$ (4) is trivial.

  (4) $\implies$ (1). If $x \in S^{\perp}$, we have $\|x\|^2 = 0$, which shows $S^{\perp} = \{0\}$. $\qquad\square$

The expansion in (2) is called the *Fourier expansion*. The equality (4) is called the *Parseval identity*, and (3) is sometimes called the generalized Parseval identity.

Corollary 2.4. *Every separable infinite-dimensional Hilbert space is isometrically isomorphic to ℓ^2.*

Proof. Let $\mathcal{H}$ be a separable infinite-dimensional Hilbert space. We fix a CONS $\{e_k\}_{k=1}^{\infty}$ of $\mathcal{H}$. Then, the map

$$\ell^2 \ni (a_k)_{k=1}^{\infty} \mapsto \sum_{k=1}^{\infty} a_k e_k \in \mathcal{H}$$

is a surjective isometry. $\qquad\square$

2.4 Bounded Operators on Hilbert Spaces

Throughout this section, the symbols $\mathcal{H}$ and $\mathcal{H}_i$ stand for Hilbert spaces.

We generalize the definition of a sesquilinear form introduced in Section 2.1 to a map $f : \mathcal{H}_1 \times \mathcal{H}_2 \to \mathbb{C}$. Namely, we say that f is a sesquilinear form if the maps $x \mapsto f(x, y)$ and $y \mapsto \overline{f(x, y)}$ are linear.

For $T \in \mathbf{B}(\mathcal{H}_1, \mathcal{H}_2)$, let $f_T(x, y) = \langle Tx, y \rangle$. Then, f_T is a sesquilinear form, and all information of T is encoded in f_T. First we note that if $S, T \in \mathbf{B}(\mathcal{H}_1, \mathcal{H}_2)$ and $f_S = f_T$, we have $S = T$. Indeed, in this case, we have $f_{S-T} = 0$ and

$$0 = f_{S-T}(x, (S - T)x) = \|(S - T)x\|^2,$$

which shows $S = T$. The definition of the operator norm shows

$$\|T\| = \sup_{\|x\| \leq 1} \|Tx\| = \sup_{\|x\| \leq 1} \sup_{\|y\| \leq 1} |\langle Tx, y \rangle| = \sup_{(x,y) \in B_{\mathcal{H}_1} \times B_{\mathcal{H}_2}} |f_T(x, y)|.$$

We define the norm of a general sesquilinear form $f : \mathcal{H}_1 \times \mathcal{H}_2 \to \mathbb{C}$ by

$$\|f\| = \sup_{(x,y) \in B_{\mathcal{H}_1} \times B_{\mathcal{H}_2}} |f(x, y)|.$$

When $\|f\|$ is finite, we say that f is *bounded*.

Lemma 2.5. *For every bounded sesquilinear form $f : \mathcal{H}_1 \times \mathcal{H}_2 \to \mathbb{C}$, there exists unique $T \in \mathbf{B}(\mathcal{H}_1, \mathcal{H}_2)$ satisfying $f = f_T$. Moreover, we have $\|f\| = \|T\|$.*

Proof. Since we have already shown the second part, we show the existence of T. We fix $x \in \mathcal{H}_1$ and set $\psi_x(y) = \overline{f(x, y)}$. Then, ψ_x is a linear functional on $\mathcal{H}_2$. From $|\psi_x(y)| \leq \|f\| \|x\| \|y\|$, we have $\psi_x \in \mathcal{H}_2^*$, and the Riesz representation theorem shows that there exists unique $\tilde{x} \in \mathcal{H}_2$ satisfying $\psi_x(y) = \langle y, \tilde{x} \rangle$ for all $y \in \mathcal{H}_2$. Thus, $f(x, y) = \langle \tilde{x}, y \rangle$. The uniqueness of $\tilde{x}$ and the linearity of $x \mapsto f(x, y)$ show that the map $x \mapsto \tilde{x}$ is linear, which we denote by T. The inequality $\|Tx\|^2 = f(x, Tx) \leq \|f\| \|x\| \|Tx\|$ implies $\|Tx\| \leq \|f\| \|x\|$, which shows that T is bounded. $\square$

Theorem 2.6. *For every $T \in \mathbf{B}(\mathcal{H}_1, \mathcal{H}_2)$, there exists a unique $T^* \in \mathbf{B}(\mathcal{H}_2, \mathcal{H}_1)$ satisfying $\langle Tx, y \rangle = \langle x, T^*y \rangle$ for all $x \in \mathcal{H}_1$ and $y \in \mathcal{H}_2$. Moreover, $\|T\| = \|T^*\|$ holds.*

Proof. We define a sesquilinear form $f : \mathcal{H}_2 \times \mathcal{H}_1 \to \mathbb{C}$ by $f(y, x) = \langle y, Tx \rangle$. Then, f is bounded and $\|f\| = \|T\|$. Thus, there exists unique $T^* \in \mathbf{B}(\mathcal{H}_2, \mathcal{H}_1)$ satisfying $f = f_{T^*}$, and $\|T^*\| = \|f\| = \|T\|$ holds. $\qquad\square$

The operator T^* is called the *adjoint operator* of T. For all $S, T \in \mathbf{B}(\mathcal{H}_1, \mathcal{H}_2)$ and $\alpha \in \mathbb{C}$, we have $(T^*)^* = T$, $(S + T)^* = S^* + T^*$, $(\alpha T)^* = \overline{\alpha} T^*$. Also, for all $T \in \mathbf{B}(\mathcal{H}_1, \mathcal{H}_2)$ and $S \in \mathbf{B}(\mathcal{H}_2, \mathcal{H}_3)$, we have $(ST)^* = T^*S^*$.

Example 2.4. We regard $\mathbb{C}^n$ as a Hilbert space with the canonical inner product and identify $\mathbf{B}(\mathbb{C}^n)$ with the matrix algebra $\mathbf{M}_n(\mathbb{C})$. Then, the adjoint operator T^* of $T \in \mathbf{M}_n(\mathbb{C})$ is nothing but the Hermitian conjugate of T.

Example 2.5. Let U be the bilateral shift operator on $\ell^2(\mathbb{Z})$. Since

$$\langle Ua, b \rangle = \sum_{k \in \mathbb{Z}} a_{k-1}\overline{b_k} = \sum_{k \in \mathbb{Z}} a_k \overline{b_{k+1}}$$

holds for all $a, b \in \ell^2(\mathbb{Z})$, we get $(U^*b)_k = b_{k+1}$.

Proposition 2.2. *For $T \in \mathbf{B}(\mathcal{H}_1, \mathcal{H}_2)$, the following hold:*

(1) $\mathcal{R}(T)^\perp = \ker T^*$.
(2) $\|T^*T\| = \|T\|^2$.

Proof. (1) follows from

$$y \in \mathcal{R}(T)^\perp \iff \forall x \in \mathcal{H}_1, \ 0 = \langle Tx, y \rangle = \langle x, T^*y \rangle.$$

(2) The submultiplicativity of the operator norm shows $\|T^*T\| \leq \|T^*\|\|T\| = \|T\|^2$. On the other hand,

$$\|Tx\|^2 = \langle Tx, Tx \rangle = \langle T^*Tx, x \rangle \leq \|T^*Tx\|\|x\| \leq \|T^*T\|\|x\|^2$$

implies $\|T\|^2 \leq \|T^*T\|$. $\qquad\square$

(2) is called the $\mathbf{C}^*$-*condition*.

Problem 2.3. Let V be the unilateral shift operator on ℓ^2.

(1) Compute the adjoint operator V^* of V.
(2) Show that $\mathcal{R}(I - V)$ is dense in ℓ^2.

(3) For $\alpha \in \mathbb{R}$, we define $W_\alpha \in \mathbf{B}(\ell^2)$ by

$$(W_\alpha a)_k = \begin{cases} 0, & k = 1, \\ \left(\dfrac{k}{k-1}\right)^\alpha a_{k-1}, & k \geq 2. \end{cases}$$

Determine α for which $\mathcal{R}(I - W_\alpha)$ is dense in ℓ^2.

Definition 2.6. We say that $A \in \mathbf{B}(\mathcal{H})$ is *self-adjoint* if $A = A^*$. We denote by $\mathbf{B}(\mathcal{H})_{\text{sa}}$ the set of self-adjoint operators in $\mathbf{B}(\mathcal{H})$. We say that $T \in \mathbf{B}(\mathcal{H})$ is *normal* if $T^*T = TT^*$ holds. Every self-adjoint operator is normal.

For $T \in \mathbf{B}(\mathcal{H})$, let $A = 2^{-1}(T + T^*)$ and $B = (2i)^{-1}(T - T^*)$. Then, $A, B \in \mathbf{B}(\mathcal{H})_{\text{sa}}$ and $T = A + iB$. This is called the *Cartesian decomposition* of T, and A and B are called the real and imaginary parts of T, respectively. The operator T is normal if and only if $AB = BA$.

Example 2.6. Let (Ω, μ) be a measure space. For $f \in L^\infty(\Omega, \mu)$, let $M_f \in \mathbf{B}(L^2(\Omega, \mu))$ be the multiplication operator of f. For any $g, h \in L^2(\Omega, \mu)$, we have

$$\langle M_f g, h \rangle = \int_\Omega f(\omega)g(\omega)\overline{h(\omega)}d\mu(\omega) = \int_\Omega g(\omega)\overline{\overline{f(\omega)}h(\omega)}d\mu(\omega)$$

$$= \langle g, M_{\bar{f}} h \rangle,$$

and $M_f^* = M_{\bar{f}}$. From the definition of the multiplication operators, we have $M_{f_1} M_{f_2} = M_{f_1 f_2}$, and M_f is a normal operator. For real-valued f, it is self-adjoint.

Note that $S, T \in \mathbf{B}(\mathcal{H})$, with $\langle Sx, x \rangle = \langle Tx, x \rangle$, for all $x \in \mathcal{H}$ implies $S = T$ thanks to the polarization identity.

Proposition 2.3. *For $A \in \mathbf{B}(\mathcal{H})$, the following conditions are equivalent:*

(1) $A \in \mathbf{B}(\mathcal{H})_{\text{sa}}$.
(2) $\langle Ax, x \rangle \in \mathbb{R}$ *for all $x \in \mathcal{H}$.*

Proof. Since

$$\langle A^*x, x \rangle = \langle x, Ax \rangle = \overline{\langle Ax, x \rangle}$$

holds for all $A \in \mathbf{B}(\mathcal{H})$ and $x \in \mathcal{H}$, (1) and (2) are equivalent. $\square$

Theorem 2.7. *For $P \in \mathbf{B}(\mathcal{H})$, the following conditions are equivalent:*

(1) *P is the projection operator onto a closed subspace of $\mathcal{H}$.*
(2) *$P = P^2 = P^*$.*

Proof. (1) $\implies$ (2). Let $\mathcal{K}$ be a closed subspace of $\mathcal{H}$, and let $P = P_{\mathcal{K}}$. Then, we have $P^2 = P$. Let $x = x_0 + x_1$ and $y = y_0 + y_1$, with $x_0, y_0 \in \mathcal{K}$, $x_1, y_1 \in \mathcal{K}^\perp$. Then,

$$\langle Px, y \rangle = \langle x_0, y_0 \rangle = \langle x, Py \rangle,$$

and $P \in \mathbf{B}(\mathcal{H})_{\mathrm{sa}}$.

(2) $\implies$ (1). We first note that $\mathcal{K} := \mathcal{R}(P)$ is closed. Indeed, if $\{a_n\}_{n=1}^\infty$ is a sequence in $\mathcal{R}(P)$ converging to $a \in \mathcal{H}$, as $P^2 = P$ implies $a_n = Pa_n$, we get $a = Pa$. Since

$$\langle Px, (I - P)y \rangle = \langle (I - P^*)Px, y \rangle = \langle (P - P^2)x, y \rangle = 0,$$

we get $\mathcal{R}(P) \perp \mathcal{R}(I - P)$. Since $x = Px + (I - P)x$, the uniqueness of orthogonal decomposition shows that $P = P_{\mathcal{K}}$. $\square$

We call an operator satisfying the above equivalent conditions a *projection operator* (without referring to the corresponding subspace) or simply a *projection*. We denote by $\mathcal{P}(\mathcal{H})$ the set of projection operators in $\mathbf{B}(\mathcal{H})$.

Proposition 2.4. *For $V \in \mathbf{B}(\mathcal{H}_1, \mathcal{H}_2)$, the following conditions are equivalent:*

(1) *For any $x, y \in \mathcal{H}_1$, we have $\langle Vx, Vy \rangle = \langle x, y \rangle$.*
(2) *$V^*V = I$.*
(3) *V is an isometry.*

Proof. Since $\langle Vx, Vy \rangle = \langle V^*Vx, y \rangle$, (1) and (2) are equivalent. (1) $\implies$ (3) is trivial. (3) $\implies$ (1) follows from the polarization identity. $\square$

Definition 2.7. We say that $U \in \mathbf{B}(\mathcal{H}_1, \mathcal{H}_2)$ is a *unitary operator* if $U^*U = I_{\mathcal{H}_1}$ and $UU^* = I_{\mathcal{H}_2}$ hold. This condition is equivalent to that U and U^* are isometries, which is further equivalent to that U is a surjective isometry. We denote the set of unitary operators from $\mathcal{H}_1$ to $\mathcal{H}_2$ by $\mathcal{U}(\mathcal{H}_1, \mathcal{H}_2)$, and we write $\mathcal{U}(\mathcal{H}, \mathcal{H}) = \mathcal{U}(\mathcal{H})$. Every element in $\mathcal{U}(\mathcal{H})$ is a normal operator.

Example 2.7.

(1) The bilateral shift on $\ell^2(\mathbb{Z})$ is a unitary operator.
(2) The unilateral shift V on ℓ^2 is an isometry but not a unitary operator. In fact, $I - VV^*$ is the projection onto $\mathbb{C}\delta_1$.

Definition 2.8. We say that $S \in \mathbf{B}(\mathcal{H}_1)$ and $T \in \mathbf{B}(\mathcal{H}_2)$ are *unitarily equivalent* if there exists a unitary $U \in \mathcal{U}(\mathcal{H}_1, \mathcal{H}_2)$ satisfying $TU = US$.

Two unitarily equivalent operators share the same properties.

Example 2.8. Let $\mathbb{T} = \mathbb{R}/2\pi\mathbb{Z}$, and let

$$L^2(\mathbb{T}) = L^2\left(\mathbb{T}, \frac{dt}{2\pi}\right).$$

For $n \in \mathbb{Z}$, let $e_n(t) = e^{int}$. Then, $\{e_n\}_{n\in\mathbb{Z}}$ is an ONS in $L^2(\mathbb{T})$. Since the trigonometric polynomials are dense in $L^2(\mathbb{T})$, it is a CONS for $L^2(\mathbb{T})$. For $f \in L^2(\mathbb{T})$ and $n \in \mathbb{Z}$, we define the *Fourier coefficient* by

$$\hat{f}(n) = \langle f, e_n \rangle = \frac{1}{2\pi} \int_0^{2\pi} f(t)e^{-int}dt.$$

Then,

$$f = \sum_{n\in\mathbb{Z}} \hat{f}(n)e_n$$

holds in $L^2(\mathbb{T})$, and the Parseval identity shows

$$\frac{1}{2\pi} \int_0^{2\pi} |f(t)|^2 dt = \sum_{n\in\mathbb{Z}} |\hat{f}(n)|^2.$$

The operator

$$F : L^2(\mathbb{T}) \ni f \mapsto \hat{f} \in \ell^2(\mathbb{Z})$$

is a surjective isometry and hence a unitary operator.

Problem 2.4. Show that the multiplication operator $M_{e_1} \in \mathcal{U}(L^2(\mathbb{T}))$ of $e_1(t) = e^{it}$ and the bilateral shift U on $\ell^2(\mathbb{Z})$ are unitarily equivalent.

Exercises

Exercise 2.1
Let P_+ be the projection from $L^2(\mathbb{T})$ onto $H^2(\mathbb{T}) := \overline{\mathrm{span}\{e_k\}_{k=0}^{\infty}}$. For $f \in L^\infty(\mathbb{T})$, we define $T_f \in \mathbf{B}(H^2(\mathbb{T}))$ by $T_f h = P_+ f h$. Show that T_{e_1} is unitarily equivalent to the unilateral shift on ℓ^2. We call $H^2(\mathbb{T})$ the *Hardy space* and T_f the *Toeplitz operator* with *symbol* f.

Exercise 2.2
For a finite measure space (Ω, μ), show $L^2(\Omega, \mu) \subset L^1(\Omega, \mu)$ and $\|f\|_1 \leq \sqrt{\mu(\Omega)}\|f\|_2$ for all $f \in L^2(\Omega, \mu)$.

Exercise 2.3
For $k \in C[0,1]^2$ and $f \in L^2[0,1]$, let $A_k f(s) = \int_0^1 k(s,t) f(t) dt$. Show $A_k \in \mathbf{B}(L^2[0,1])$, and compute its adjoint operator $A_k{}^*$.

Exercise 2.4
Let $H(\mathbb{D})$ be the set of holomorphic functions on the unit disc $\mathbb{D}$.

(1) Let $z \in \mathbb{D}$, and let $0 < r < d(z, \partial\mathbb{D})$. Show that the following holds for all $f \in H(\mathbb{D})$:

$$|f(z)| \leq \frac{1}{\sqrt{\pi}r}\sqrt{\int_{B(z,r)} |f(x+iy)|^2 dx dy}.$$

(2) Show that $A^2(\mathbb{D}) := H(\mathbb{D}) \cap L^2(\mathbb{D})$ is a closed subspace of $L^2(\mathbb{D})$. We call $A^2(\mathbb{D})$ the *Bergman space*.

(3) Show that

$$\left\{ \sqrt{\frac{n+1}{\pi}} z^n \right\}_{n=0}^{\infty}$$

is a CONS for $A^2(\mathbb{D})$.

(4) We define $W \in \mathbf{B}(A^2(\mathbb{D}))$ by $Wf(z) = zf(z)$. Show that W is unitarily equivalent to $W_{-1/2}$ in Problem 2.3.

(5) For $w \in \mathbb{D}$, we define $k_w \in A^2(\mathbb{D})$ by

$$k_w(z) = \sum_{n=0}^{\infty} \sqrt{\frac{n+1}{\pi}} z^n \sqrt{\frac{n+1}{\pi}} \overline{w}^n = \frac{1}{\pi} \frac{1}{(1 - \overline{w}z)^2}.$$

Show $\langle f, k_w \rangle = f(w)$ for all $f \in A^2(\mathbb{D})$. The function $K(z, w) = k_w(z)$ is called the *Bergman kernel*.

Exercise 2.5

Assume that the Fourier transform $\hat{f}(\xi)$ of $f \in L^2(\mathbb{R})$ satisfies $\hat{f}(\xi) \neq 0$ almost everywhere. For $r \in \mathbb{R}$, let $f_r(t) = f(t - r)$. Show that span$\{f_r\}_{r \in \mathbb{R}}$ is dense in $L^2(\mathbb{R})$.

Exercise 2.6

Let $(\Omega, \mathcal{F})$ be a measurable space, and let $\mu, \nu : \mathcal{F} \to [0, \infty]$ be σ-finite measures. If $\mu(E) = 0$ implies $\nu(E) = 0$ for every $E \in \mathcal{F}$, we say that ν is absolutely continuous with respect to μ and write $\nu \ll \mu$. When $\mu \ll \nu$ and $\nu \ll \mu$ hold, we say that μ and ν are equivalent. In this exercise, as an application of the Riesz representation theorem, we deduce the following Radon–Nikodym theorem: If $\nu \ll \mu$, there exists a measurable function $h : \Omega \to [0, \infty)$ such that $\nu(E) = \int_E h d\mu$ for every $E \in \mathcal{F}$. The function h is unique up to a null set with respect to μ; it is denoted by $d\nu/d\mu$ and called the *Radon–Nikodym derivative*.

We consider only the case where μ and ν are finite measures in the following. Let $m = \mu + \nu$, and let $\mathcal{H} = L^2(\Omega, m)$.

(1) Define linear functions $\varphi, \psi : \mathcal{H} \to \mathbb{C}$ by $\varphi(f) = \int_\Omega f d\mu$ and $\psi(f) = \int_\Omega f d\nu$, respectively. Show that φ and ψ are bounded.

(2) The Riesz representation theorem and (1) show that there exist unique $h_\varphi, h_\psi \in \mathcal{H}$ such that $\varphi(f) = \langle f, h_\varphi \rangle_{\mathcal{H}}$ and $\psi(f) = \langle f, h_\psi \rangle_{\mathcal{H}}$ for every $f \in \mathcal{H}$. Show $0 \le h_\varphi, h_\psi \le 1$, $h_\varphi + h_\psi = 1$, and $\mu\{h_\varphi = 0\} = 0$.

(3) Prove the Radon–Nikodym theorem for finite measures.

Exercise 2.7

Let (Ω, μ) be a σ-finite measure space.

(1) Show that there exists a measurable function $\rho : \Omega \to (0, \infty)$ such that the measure defined by $\mu_0(E) = \int_E \rho d\mu$ is a finite measure.

(2) Prove the Radon–Nikodym theorem for σ-finite measures.

Chapter 3

The Dual Spaces

Duality is a fundamental idea in functional analysis. In this chapter, we show the Hahn–Banach extension theorem, which guarantees that the dual space of a Banach space (or, more generally, a locally convex space) is sufficiently large, and we give its applications to Banach space theory.

3.1   The Hahn–Banach Extension Theorem

In the following theorem, we make the only exception in this book dealing with real linear spaces.

Theorem 3.1 (Hahn–Banach extension theorem (real version)). *Let X be a real linear space, and assume that a function $p : X \to \mathbb{R}$ satisfies the following conditions:*

(1) $p(x + y) \le p(x) + p(y)$ *for all* $x, y \in X$.
(2) $p(tx) = tp(x)$ *for all* $x \in X$ *and* $t > 0$.

Let X_0 be a subspace of X, and assume that a linear functional $\psi_0 : X_0 \to \mathbb{R}$ satisfies $\psi_0(x) \le p(x)$ for all $x \in X_0$. Then, there exists a linear functional $\psi : X \to \mathbb{R}$ satisfying $\psi|_{X_0} = \psi_0$ and $\psi(x) \le p(x)$ for all $x \in X$.

Proof.   We assume $X \ne X_0$, as there is nothing to show if $X = X_0$. First, we show that ψ_0 extends to a linear functional satisfying the above condition on a subspace containing X_0 as a subspace of codimension 1. We take $x_1 \in X \backslash X_0$ and set $X_1 = X_0 + \mathbb{R}x_1$.

As $X_1 = X_0 \oplus \mathbb{R}x_1$, for every real number α, there exists a linear functional $\varphi_\alpha : X_1 \to \mathbb{R}$ satisfying $\varphi_\alpha|_{X_0} = \psi_0$ and $\varphi_\alpha(x_1) = \alpha$. We show that φ_α satisfies the above condition if we choose appropriate α. For $x_0 \in X_0$ and $t > 0$, we have

$$p(x_0 + tx_1) - \varphi_\alpha(x_0 + tx_1) = t \left(p \left(\frac{1}{t}x_0 + x_1 \right) - \psi_0 \left(\frac{1}{t}x_0 \right) - \alpha \right),$$

$$p(x_0 - tx_1) - \varphi_\alpha(x_0 - tx_1) = t \left(p \left(\frac{1}{t}x_0 - x_1 \right) - \psi_0 \left(\frac{1}{t}x_0 \right) + \alpha \right).$$

Thus, if $\gamma \le \beta$ holds for

$$\beta = \inf\{p(x + x_1) - \psi_0(x); \ x \in X_0\},$$
$$\gamma = \sup\{\psi_0(x) - p(x - x_1); \ x \in X_0\},$$

we can choose α, with $\gamma \le \alpha \le \beta$, so that φ_α satisfies the condition. In fact, for all $x, y \in X_0$, we have

$$(p(x + x_1) - \psi_0(x)) - (\psi_0(y) - p(y - x_1))$$
$$= p(x + x_1) + p(y - x_1) - \psi_0(x + y) \ge p(x + y) - \psi_0(x + y) \ge 0,$$

and $\gamma \le \beta$ holds.

Now, we prove the theorem by using Zorn's lemma and the fact we have just shown. Let $\mathcal{E}$ be the set of pairs (Y, ψ) satisfying the following conditions: Y is a subspace of X containing X_0 and $\psi : Y \to \mathbb{R}$ is a linear functional satisfying $\psi|_{X_0} = \psi_0$ and $\psi(y) \le p(y)$ for all $y \in Y$. By setting $(Y_1, \psi_1) \le (Y_2, \psi_2)$, if $(Y_1, \psi_1), (Y_2, \psi_2) \in \mathcal{E}$ satisfy $Y_1 \subset Y_2$ and $\psi_2|_{Y_1} = \psi_1$, we can make $\mathcal{E}$ an ordered set.

We show that if $\mathcal{E}_0 \subset \mathcal{E}$ is a totally ordered subset, it has an upper bound. Let $Y_0 = \bigcup_{(Y,\psi) \in \mathcal{E}_0} Y$. Then, Y_0 is closed under scalar multiplication. Let $x_1, x_2 \in Y_0$. Then, there exist $(Y_i, \psi_i) \in \mathcal{E}_0$, $i = 1, 2$, satisfying $x_i \in Y_i$. Since $\mathcal{E}_0$ is totally ordered, either $(Y_1, \psi_1) \le (Y_2, \psi_2)$ or $(Y_2, \psi_2) \le (Y_1, \psi_1)$ holds. In either case, we have $x + y \in Y_0$, and Y_0 is a subspace of X. We would like to define $\psi_0'(x)$ for $x \in Y_0$ by choosing $(Y, \psi) \in \mathcal{E}_0$, with $x \in Y$, and setting $\psi_0'(x) = \psi(x)$. Again since $\mathcal{E}_0$ is totally ordered, the number $\psi(x)$ does not depend on the choice of (Y, ψ), and ψ_0' is uniquely determined. By construction, we have $(Y_0, \psi_0') \in \mathcal{E}$, and it is an upper bound of $\mathcal{E}_0$.

By Zorn's lemma, there exists a maximal element(Z, ψ) in $\mathcal{E}$. If $Z \neq X$, we could further extend ψ satisfying the condition, which contradicts the maximality of (Z, ψ). Thus, ψ is the desired extension. $\qquad\square$

Let X be a complex linear space, which is a real linear space as well. Then, the complex linear functionals on X and the real linear functionals on X have the same information. Indeed, if $\varphi : X \to \mathbb{C}$ is a complex linear functional, its real part $\operatorname{Re}\varphi : X \to \mathbb{R}$ is a real linear functional. As $\operatorname{Re}\varphi(ix) = -\operatorname{Im}\varphi(x)$, we can recover φ from $\operatorname{Re}\varphi$ as

$$\varphi(x) = \operatorname{Re}\varphi(x) - i\operatorname{Re}\varphi(ix).$$

Problem 3.1. Let X be a complex linear space, and let $\psi : X \to \mathbb{R}$ be a real linear functional on X. Show that if we define $\varphi : X \to \mathbb{C}$ by

$$\varphi(x) = \psi(x) - i\psi(ix),$$

then φ is a complex linear functional.

Definition 3.1. Let X be a complex linear space. We say that $p : X \to [0, \infty)$ is a *semi-norm* if the following hold:

(1) $p(\alpha x) = |\alpha|p(x)$ for all $x \in X$ and $\alpha \in \mathbb{C}$.
(2) $p(x + y) \leq p(x) + p(y)$ for all $x, y \in X$.

Theorem 3.2 (Hahn–Banach extension theorem (complex version)). *Let X be a complex linear space, let $p : X \to [0, \infty)$ be a semi-norm, let X_0 be a (complex) subspace of X, and let $\varphi_0 : X_0 \to \mathbb{C}$ be a complex linear functional such that $|\varphi_0(x_0)| \leq p(x_0)$ holds for all $x_0 \in X_0$. Then, there exists a complex linear functional $\varphi : X \to \mathbb{C}$ satisfying $\varphi|_{X_0} = \varphi_0$ and $|\varphi(x)| \leq p(x)$ for all $x \in X$.*

Proof. Applying the previous theorem to the real linear functional $\psi_0 = \operatorname{Re}\varphi_0 : X_0 \to \mathbb{R}$, we can get a real linear functional $\psi : X \to \mathbb{R}$ with $\psi|_{X_0} = \operatorname{Re}\varphi_0$ satisfying $\psi(x) \leq p(x)$ for all $x \in X$. Letting $\varphi(x) = \psi(x) - i\psi(ix)$, we get a complex linear functional φ, with $\varphi|_{X_0} = \varphi_0$. For $x \in X$, we choose a complex number ω with modulus 1 satisfying $\omega\varphi(x) = |\varphi(x)|$. Then, $|\varphi(x)| = \psi(\omega x) \leq p(\omega x) = p(x)$. $\qquad\square$

In the remainder of this book, the word "linear" always means "complex linear" again.

As we saw above, there is no mention of the topology of X in the statement of the Hahn–Banach extension theorem. For actual applications, the information of the topology is carried by $p(x)$.

Corollary 3.1. *Let X be a normed space, let X_0 be a subspace of X, and let $\varphi_0 \in X_0^*$. Then, there exists an extension $\varphi \in X^*$ of φ_0 satisfying $\|\varphi\| = \|\varphi_0\|$.*

Proof. Taking $p(x) = \|\varphi_0\|\|x\|$ as a semi-norm, we can extend φ_0 by applying the Hahn–Banach extension theorem. $\qquad\square$

Corollary 3.2. *Let X be a normed space. Then, for every $a \in X$, there exists $\varphi \in X^*$ satisfying $\|\varphi\| = 1$ and $\varphi(a) = \|a\|$.*

Proof. Taking $\|x\|$ as a semi-norm, we can apply the Hahn–Banach extension theorem to $X_0 = \mathbb{C}a$ and $\varphi_0(\alpha a) = \alpha\|a\|$. $\qquad\square$

Let X be a normed space. We denote $(X^*)^*$ by X^{**} and call it the *second dual* of X. For $x \in X$, if we define $x^{**} \in X^{**}$ by $x^{**}(\varphi) = \varphi(x)$, we have

$$|x^{**}(\varphi)| = |\varphi(x)| \le \|\varphi\|\|x\|,$$

and $\|x^{**}\| \le \|x\|$. On the other hand, there exists $\varphi \in X^*$ satisfying $\|\varphi\| = 1$ and $\varphi(x) = \|x\|$, and so $\|x^{**}\| = \|x\|$.

Corollary 3.3. *For a normed space X, the map $X \ni x \mapsto x^{**} \in X^{**}$ is an isometry.*

In what follows, we identify x with x^{**}, and we regard X as a subspace of X^{**}.

Definition 3.2. We say that a Banach space X is *reflexive* if $X = X^{**}$ holds.

Finite-dimensional spaces and Hilbert spaces are examples of reflexive Banach spaces.

Theorem 3.3. *Let Y be a closed subspace of a Banach space X, let $Q : X \to X/Y$ be the quotient map, and let $Y^\perp = \{\varphi \in X^*; \forall y \in Y, \varphi(y) = 0\}$. For $\varphi \in X^*$, we denote $\varphi + Y^\perp \in X^*/Y^\perp$ by $[\varphi]$.*

(1) *The map* $\Phi : X^*/Y^\perp \to Y^*$, $[\varphi] \mapsto \varphi|_Y$ *is an isometric isomorphism.*
(2) *The map* $\Psi : (X/Y)^* \to Y^\perp$, $\psi \mapsto \psi \circ Q$ *is an isometric isomorphism.*

Proof. We leave (2) for the reader, as it can be shown using the knowledge in Chapter 1, and we show only (1) when $Y \neq \{0\}$. Since the kernel of the restriction map $\rho : X^* \to Y^*$, $\varphi \mapsto \varphi|_Y$ is $Y^\perp$, Theorem 1.5 implies that Φ is injective and $\|\Phi\| = \|\rho\| \leq 1$. Let $\psi \in Y^*$. Then, there exists an extension $\varphi \in X^*$ of ψ satisfying $\|\varphi\| = \|\psi\|$. Thus, Φ is surjective and $\|\Phi([\varphi])\| = \|\varphi\| \geq \|[\varphi]\|$. Therefore, Φ is an isometric isomorphism. $\qquad\square$

Problem 3.2. Show (2) in the above theorem.

3.2 Examples of Dual Spaces

3.2.1 *Hilbert spaces*

The following claim requires a little caution: The Riesz representation theorem shows that the dual space of a Hilbert space $\mathcal{H}$ is $\mathcal{H}$. For $y \in \mathcal{H}$, we define $\varphi_y \in \mathcal{H}^*$ by $\varphi_y(x) = \langle x, y \rangle$. Then, the map $y \mapsto \varphi_y$ is not linear but conjugate linear. The remedy for this problem is to introduce the notion of the complex conjugate Hilbert space $\overline{\mathcal{H}}$ of $\mathcal{H}$. Let $\overline{\mathcal{H}} = \{\overline{x}; \ x \in \mathcal{H}\}$. Namely, $\overline{\mathcal{H}}$ is $\mathcal{H}$ as a set and as a module as well. We introduce a scalar multiplication into $\overline{\mathcal{H}}$ by $\alpha \cdot \overline{x} = \overline{\overline{\alpha}x}$ and an inner product by $\langle \overline{x}, \overline{y} \rangle_{\overline{\mathcal{H}}} = \langle y, x \rangle$. Then, $\overline{\mathcal{H}}$ is a Hilbert space. The Riesz representation theorem shows that the map $\overline{x} \mapsto \varphi_x$ induces an isometric isomorphism from $\overline{\mathcal{H}}$ onto $\mathcal{H}^*$.

Problem 3.3. Show that a Hilbert space is reflexive.

3.2.2 *L^p spaces*

Let (Ω, μ) be a σ-finite measure space, and let $1 \leq p, q \leq \infty$, with $1/p + 1/q = 1$. The Hölder inequality (see Theorem A.1) shows that for $g \in L^q(\Omega, \mu)$, we can define $\varphi_g \in L^p(\Omega, \mu)^*$ by $\varphi_g(f) = \int_\Omega f(\omega)g(\omega)d\mu(\omega)$, and $\|\varphi_g\| \leq \|g\|_q$ holds.

Lemma 3.1. *The map* $\Phi : L^q(\Omega, \mu) \to L^p(\Omega, \mu)^*$, $g \mapsto \varphi_g$, *is an isometry.*

Proof. As we already know $\|\varphi_g\| \leq \|g\|_q$, we only show $\|\varphi_g\| \geq \|g\|_q$, assuming $\|g\|_q \neq 0$.

Case $p = 1$. For $\|g\|_\infty > \lambda > 0$, the set $E = \{|g| > \lambda\}$ satisfies $\mu(E) \neq 0$. Since (Ω, μ) is σ-finite, there exists an increasing sequence of measurable sets $\Omega_1 \subset \Omega_2 \subset \cdots$ such that $\mu(\Omega_n) < \infty$ and $\bigcup_{n=1}^\infty \Omega_n = \Omega$. Let $E_n = E \cap \Omega_n$. Since $\lim_{n\to\infty} \mu(E_n) = \mu(E)$, there exists $N \in \mathbb{N}$ with $0 < \mu(E_N) < \infty$. Let

$$
f(\omega) = \begin{cases} \dfrac{1}{\mu(E_N)} \dfrac{\overline{g(\omega)}}{|g(\omega)|}, & \omega \in E_N, \\[2mm] 0, & \omega \in \Omega \backslash E_N. \end{cases}
$$

Then, we have $\|f\|_1 = 1$, and

$$
\varphi_g(f) = \frac{1}{\mu(E_N)} \int_{E_N} |g(\omega)| d\mu(\omega) \geq \lambda.
$$

As $\|g\|_\infty > \lambda > 0$ is arbitrary, we get $\|\varphi_g\| \geq \|g\|_\infty$.

Case $1 < p < \infty$. Let

$$
f(\omega) = \begin{cases} |g(\omega)|^{\frac{q}{p}} \dfrac{\overline{g(\omega)}}{|g(\omega)|}, & g(\omega) \neq 0, \\[2mm] 0, & g(\omega) = 0. \end{cases}
$$

Then, we have $\|f\|_p = \|g\|_q^{\frac{q}{p}}$, and

$$
\varphi_g(f) = \int_\Omega |g(\omega)|^{1 + \frac{q}{p}} d\mu(\omega) = \int_\Omega |g(\omega)|^q d\mu(\omega) = \|g\|_q^q = \|g\|_q^{1 + \frac{q}{p}},
$$

which shows $\|\varphi_g\| \geq \|g\|_q$. The case of $p = \infty$ can be treated in the same way. $\qquad\qquad\qquad\qquad\qquad\qquad\qquad\qquad\qquad\qquad\qquad\qquad\square$

Theorem 3.4. *When $1 \leq p < \infty$, the map $\Phi : L^q(\Omega, \mu) \to L^p(\Omega, \mu)^*$ is an isometric isomorphism.*

Proof. We give the proof only in the case of sequence spaces here. For the general case, the reader is referred to Exercise 3.2 in the case of $p = 1$, and Section A.3 in the case of $1 < p < \infty$.

Since $1 \leq p < \infty$, the subspace $c_{0,0} = \mathrm{span}\{\delta_n\}_{n=1}^\infty$ is dense in ℓ^p. Indeed, for every $a \in \ell^p$, we have

$$\left\| a - \sum_{k=1}^n a_k \delta_k \right\|_p = \left(\sum_{k=n+1}^\infty |a_k|^p \right)^{\frac{1}{p}} \to 0, \quad (n \to \infty).$$

For $\varphi \in \ell^{p*}$, let $b_k = \varphi(\delta_k)$, and define a sequence b by $b = (b_k)_{k=1}^\infty$. It suffices to show $b \in \ell^q$ and $\varphi = \varphi_b$. For $N \in \mathbb{N}$, we define $b^N \in c_{0,0}$ by $b^N = \sum_{k=1}^N b_k \delta_k$. For $a \in \mathrm{span}\{\delta_k\}_{k=1}^N$, we have

$$\varphi(a) = \varphi\left(\sum_{k=1}^N a_k \delta_k \right) = \sum_{k=1}^N a_k b_k,$$

and $|\varphi(a)| \leq \|\varphi\| \|a\|_p$. Thus, $\|b^N\|_q \leq \|\varphi\|$ holds. As this is the case for every $N \in \mathbb{N}$, we get $b \in \ell^q$. Since $\varphi, \varphi_b \in \ell^{p*}$ and they coincide on the dense subspace $c_{0,0}$, we conclude that $\varphi = \varphi_b$. $\qquad\square$

Corollary 3.4. $L^p(\Omega, \mu)$ *is reflexive for $1 < p < \infty$.*

Problem 3.4. Show that the dual space of the sequence space c_0 (see Problem 1.1) is isometrically isomorphic to ℓ^1.

As $c_0^{**} = \ell^{1*} = \ell^\infty$, the space c_0 is not reflexive. It follows that ℓ^1 is not reflexive either (see Corollary 5.5), but we directly show this fact by using only separability here.

Lemma 3.2. *For $1 \leq p < \infty$, the Banach space ℓ^p is separable, while ℓ^∞ is not.*

Proof. In the case of $1 \leq p < \infty$, the set of linear combinations of $\{\delta_n\}_{n=1}^\infty$ with coefficients in $\mathbb{Q} + i\mathbb{Q}$ is dense in ℓ^p. Thus, ℓ^p is separable.

In the case of $p = \infty$, let $2^\mathbb{N}$ be the power set of $\mathbb{N}$. For $F \in 2^\mathbb{N}$, we have $\chi_F \in \ell^\infty$, and if $F \neq F'$, we get $\|\chi_F - \chi_{F'}\|_\infty = 1$. As $2^\mathbb{N}$ is uncountable, we see that ℓ^∞ is non-separable. $\qquad\square$

Lemma 3.3. *Let X be a normed space. If X^* is separable, so is X.*

Proof. Assume that X^* is separable, and choose a dense sequence $\{\varphi_n\}_{n=1}^\infty$ in the unit sphere S_{X^*} of X^*. We choose $x_n \in S_X$ satisfying $|\varphi_n(x_n)| \geq 1/2$, and we set $Y = \overline{\mathrm{span}\{x_n\}_{n=1}^\infty}$. Then, Y is a separable closed subspace of X. We assume that $Y \neq X$ and deduce a contradiction. As $Y \neq X$, there exists $\varphi \in S_{Y^\perp} \subset S_{X^*}$. Then, there exists $n \in \mathbb{N}$ satisfying $\|\varphi - \varphi_n\| < 1/4$. Since $\|x_n\| = 1$, we get $|\varphi(x_n) - \varphi_n(x_n)| \leq \|\varphi - \varphi_n\| \leq 1/4$. On the other hand, we have $|\varphi(x_n) - \varphi_n(x_n)| = |\varphi_n(x_n)| \geq 1/2$, which is a contradiction. Thus, X is separable. $\qquad\square$

Theorem 3.5. *The sequence space ℓ^1 is not reflexive.*

Proof. If ℓ^1 were reflexive, the second dual ℓ^{1**} would be separable, and the previous lemma shows that $\ell^{1*} = \ell^\infty$ would be separable too, which is a contradiction. Thus, ℓ^1 is not reflexive. $\qquad\square$

3.2.3 *Direct sums of Banach spaces*

Definition 3.3. Let X and Y be Banach spaces. For $1 \leq p \leq \infty$, we introduce a norm into $X \oplus Y$ by

$$\|(x, y)\|_p = \begin{cases} (\|x\|^p + \|y\|^p)^{\frac{1}{p}}, & 1 \leq p < \infty, \\ \max\{\|x\|, \|y\|\}, & p = \infty. \end{cases}$$

Then, $\|\cdot\|_p$, $1 \leq p \leq \infty$, are mutually equivalent norms, and $(X \oplus Y, \|\cdot\|_p)$ are Banach spaces. In what follows, we denote $(X \oplus Y, \|\cdot\|_p)$ by $X \oplus_p Y$ for simplicity.

Problem 3.5. Show that $(X \oplus_p Y)^*$ is isometrically isomorphic to $X^* \oplus_q Y^*$, where $1/p + 1/q = 1$.

3.2.4 *The space of continuous functions*

We end this section with an illustration of the dual space of $C(\Omega)$ for a compact Hausdorff space Ω.

Let $(\Omega, \mathcal{F})$ be a measurable space. A function $m : \mathcal{F} \to \mathbb{C}$ is said to be a *complex measure* if:

(1) $m(\emptyset) = 0$,
(2) $\{E_n\}_{n=1}^\infty \subset \mathcal{F}$ are disjoint,

$$m\left(\bigcup_{n=1}^\infty E_n\right) = \sum_{n=1}^\infty m(E_n).$$

Let m be a complex measure. For $E \in \mathcal{F}$, we set

$$|m|(E) = \sup \left\{ \sum_{k=1}^{n} |m(S_k)|; \ \{S_k\}_{k=1}^{n} \subset \mathcal{F} \text{ are disjoint and } E = \bigcup_{k=1}^{n} S_k \right\}.$$

Then, $|m|$ is a finite measure. We call $|m|$ the *total variation measure* of m. We can define the integral $\int_\Omega f \, dm$ of $f \in L^1(\Omega, |m|)$ by m, and $|\int_\Omega f \, dm| \leq \int_\Omega |f| d|m|$ holds.

Let Ω be a compact Hausdorff space, and let $\mathfrak{B}_\Omega$ be the σ-algebra of all Borel subsets of Ω. Let $m : \mathfrak{B}_\Omega \to \mathbb{C}$ be a complex measure. If $|m|$ is regular, we say that m is regular (see Section A.5 for the regularity of a Borel measure).

Theorem 3.6 (Riesz–Markov–Kakutani). *Let Ω be a compact Hausdorff space, and let $\varphi \in C(\Omega)^*$. Then, there exists a unique regular complex measure $m : \mathfrak{B}_\Omega \to \mathbb{C}$ such that $\varphi(f) = \int_\Omega f \, dm$ holds for every $f \in C(\Omega)$. Moreover, we have $\|\varphi\| = |m|(\Omega)$.*

The reader is referred to Conway (1990, Theorem C. 18) and Rudin (1987, Theorem 6.19) for the proof in the general case. Here, we treat only the case of $\Omega = [a, b]$ and show the part that follows from an elementary argument using the Hahn–Banach extension theorem. Let $\mathcal{B}^b[a, b]$ be the space of bounded Borel functions on $[a, b]$. By introducing the norm $\|f\|_\infty = \sup_{t \in [a,b]} |f(t)|$ into $\mathcal{B}^b[a, b]$, we can make it a Banach space, and $C[a, b]$ is a closed subspace of $\mathcal{B}^b[a, b]$. Applying the Hahn–Banach extension theorem to $\varphi \in C[a, b]^*$, we take an extension $\psi \in \mathcal{B}^b[a, b]^*$ of φ satisfying $\|\varphi\| = \|\psi\|$, and we define a function h on $[a, b]$ by

$$h(\lambda) = \begin{cases} 0, & \lambda = a, \\ \psi(\chi_{[a,\lambda]}), & \lambda \in (a, b]. \end{cases}$$

Now, we show that $\varphi(f) = \int_a^b f(t) dh(t)$ holds for every $f \in C[a, b]$ in the sense of the Riemann–Stieltjes integral. The reader is referred to Section A.6 for the basics of the Riemann–Stieltjes integral.

The function h is of bounded variation, and its total variation is bounded by $\|\varphi\|$. Indeed, let Δ be an arbitrary partition, $a = t_0 < t_1 < \cdots < t_n = b$, of $[a, b]$, and choose complex numbers ω_i with modulus 1 satisfying $\omega_i(h(t_i) - h(t_{i-1})) = |h(t_i) - h(t_{i-1})|$. Then, we have

$$\sum_{i=1}^{n} |h(t_i) - h(t_{i-1})| = \psi \left(\omega_1 \chi_{[a,t_1]} + \sum_{i=2}^{n} \omega_i \chi_{(t_{i-1},t_i]} \right) \leq \|\psi\| = \|\varphi\|,$$

which shows that h is of bounded variation and its total variation is bounded by $\|\varphi\|$.

We define the mesh of the partition Δ by $h(\Delta) = \max_{1 \leq i \leq n} (t_i - t_{i-1})$. For $f \in C[a,b]$, we have

$$\|f - (f(t_1)\chi_{[a,t_1]} + \sum_{i=2}^{n} f(t_i)\chi_{(t_{i-1},t_i]})\|_\infty \leq m(f, h(\Delta)),$$

where

$$m(f, \delta) = \sup\{|f(s) - f(t)|; \ s,t \in [a,b], \ |s - t| \leq \delta\}.$$

Since f is uniformly continuous on $[a,b]$, we have $\lim_{\delta \to +0} m(f, \delta) = 0$. As we have

$$\sum_{i=1}^{n} f(t_i)(h(t_i) - h(t_{i-1})) = \psi(f(t_1)\chi_{[a,t_1]} + \sum_{i=2}^{n} f(t_i)\chi_{(t_{i-1},t_i]}),$$

we get

$$\left| \varphi(f) - \sum_{i=1}^{n} f(t_i)(h(t_i) - h(t_{i-1})) \right| \leq \|\varphi\| m(f, h(\Delta)) \to 0, \quad (h(\Delta) \to 0),$$

and $\varphi(f) = \int_a^b f(t)dh(t)$.

3.3 The Weak Topology and the Weak* Topology

Let X and Y be linear spaces, and we assume that $b : X \times Y \to \mathbb{C}$ is a non-degenerate bilinear form. We call such (X, Y) *linear spaces in duality*. Recall that b is non-degenerate if and only if the following two conditions are satisfied: (1) $\forall x \in X$, $b(x, y) = 0 \implies y = 0$, and (2) $\forall y \in Y$, $b(x, y) = 0 \implies x = 0$.

Definition 3.4. For $x_0 \in X$, a finite subset $F \Subset Y$, and a positive number $\varepsilon > 0$, we set

$$U(x_0; F, \varepsilon) = \{x \in X; \ \forall y \in F, \ |b(x - x_0, y)| < \varepsilon\}.$$

We denote by $\sigma(X, Y)$ the topology of X, for which $\{U(x_0; F, \varepsilon)\}_{F \Subset Y, \varepsilon > 0}$ is a fundamental system of neighborhoods of x_0. In the same way, we define the topology $\sigma(Y, X)$ of Y.

In particular, when X is a normed space, $Y = X^*$ and $b(x, \varphi) = \varphi(x)$, we call $\sigma(X, X^*)$ the *weak topology* of X, and we call $\sigma(X^*, X)$ the *weak* topology* of X^*.

When a sequence $\{x_n\}_{n=1}$ in X converges to x in the weak topology, we write w-$\lim_{n \to \infty} x_n = x$ and call it *weak convergence*. This is equivalent to the condition that $\lim_{n \to \infty} \varphi(x_n) = \varphi(x)$ holds for all $\varphi \in X^*$.

When a sequence $\{\varphi_n\}_{n=1}^{\infty}$ in X^* converges to φ in the weak* topology, we write w*-$\lim_{n \to \infty} \varphi_n = \varphi$ and call it *weak* convergence*. This is equivalent to the condition that $\lim_{n \to \infty} \varphi_n(x) = \varphi(x)$ holds for all $x \in X$.

Remark 3.1. In general, either the weak topology or the weak* topology does not satisfy the first axiom of countability, and their topological properties cannot be characterized by the convergence of sequences. For this purpose, we need to use nets, which are a generalization of sequences (see Section A.1).

Problem 3.6. Let Y be a closed subspace of a normed space X.

(1) Show that the weak topology of Y coincides with the relative topology of the weak topology of X.
(2) Show that the weak* topology of $Y^\perp$ as the dual space of X/Y coincides with the relative topology of the weak* topology of X^*.

Example 3.1. Let $\mathcal{H}$ be a Hilbert space. By the Riesz representation theorem, a sequence $\{x_n\}_{n=1}^{\infty}$ in $\mathcal{H}$ converges to $x \in \mathcal{H}$ in the weak topology if and only if $\lim_{n \to \infty} \langle x_n, y \rangle = \langle x, y \rangle$ holds for all $y \in \mathcal{H}$. Let $\{e_n\}_{n=1}^{\infty}$ be an ONS of $\mathcal{H}$. Then, the Bessel inequality $\sum_{n=1}^{\infty} |\langle x, e_n \rangle|^2 \leq \|x\|^2$ shows that $\{\langle e_n, x \rangle\}_{n=1}^{\infty}$ converges to 0 for all $x \in \mathcal{H}$, and $\{e_n\}_{n=1}^{\infty}$ weakly converges to 0.

Example 3.2. Let $1 < p < \infty$, and define $f_n \in L^p[0,1]$ by

$$f_n(t) = \begin{cases} n^{\frac{1}{p}}, & t \in \left[0, \dfrac{1}{n}\right], \\ 0, & t \in \left(\dfrac{1}{n}, 1\right]. \end{cases}$$

Then, we have $\|f_n\|_p = 1$. Recall $L^p[0,1]^* = L^q[0,1]$, with $1/p + 1/q = 1$. For $g \in L^q[0,1]$, the Hölder inequality implies

$$\left| \int_0^1 f_n(t)g(t)dt \right| = \left| \int_0^{\frac{1}{n}} f_n(t)g(t)dt \right|$$

$$\leq \|f_n\|_p \left(\int_0^{\frac{1}{n}} |g(t)|^q dt \right)^{\frac{1}{q}} \to 0 \quad (n \to \infty).$$

Thus, $\{f_n\}_{n=1}^{\infty}$ weakly converges to 0 in $L^p[0,1]$.

Example 3.3. Let $\varphi, \varphi_n \in L^1[0,1]^*$ be $\varphi(f) = 2^{-1} \int_0^1 f(t)dt$ and

$$\varphi_n(f) = \sum_{k=0}^{n-1} \int_{\frac{2k}{2n}}^{\frac{2k+1}{2n}} f(t)dt.$$

Then, $\|\varphi\| = 1/2$ and $\|\varphi_n\| = 1$. We show that $\{\varphi_n\}_{n=1}^{\infty}$ converges to φ in the weak* topology as follows.

For $f \in C[0,1]$, we have

$$|\varphi_n(f) - \varphi(f)| = \frac{1}{2} \left| \sum_{k=0}^{n-1} \left(\int_{\frac{2k}{2n}}^{\frac{2k+1}{2n}} f(t)dt - \int_{\frac{2k+1}{2n}}^{\frac{2k+2}{2n}} f(t)dt \right) \right|$$

$$= \frac{1}{2} \left| \sum_{k=0}^{n-1} \int_{\frac{2k}{2n}}^{\frac{2k+1}{2n}} \left(f(t) - f\left(t + \frac{1}{2n}\right) \right) dt \right|$$

$$\leq \frac{1}{4} \sup_{t \in [0, 1 - \frac{1}{2n}]} \left| f(t) - f\left(t + \frac{1}{2n}\right) \right|.$$

Since f is uniformly continuous on $[0,1]$, the right-hand side converges to 0 as $n \to \infty$. Thus, $\{\varphi_n(f)\}_{n=1}^{\infty}$ converges to $\varphi(f)$.

Since $C[0,1]$ is dense in $L^1[0,1]$, for every $f \in L^1[0,1]$ and every $\varepsilon > 0$, there exists $g \in C[0,1]$ satisfying $\|f - g\|_1 < \varepsilon$. As $\{\varphi_n(g)\}_{n=1}^\infty$ converges to $\varphi(g)$, there exists $N \in \mathbb{N}$ such that $n \geq N$ implies $|\varphi_n(g) - \varphi(g)| < \varepsilon$. Thus, if $n \geq N$, we get

$$
\begin{aligned}
|\varphi_n(f) - \varphi(f)| &\leq |\varphi_n(f - g)| + |\varphi_n(g) - \varphi(g)| + |\varphi(g - f)| \\
&\leq (\|\varphi_n\| + \|\varphi\|)\|f - g\|_1 + \varepsilon \\
&\leq \frac{5}{2}\varepsilon,
\end{aligned}
$$

and $\{\varphi_n(f)\}_{n=1}^\infty$ converges to $\varphi(f)$.

Note that if we identify $L^1[0,1]^*$ with $L^\infty[0,1]$, the functional φ corresponds to the constant function $1/2$, and φ_n corresponds to χ_{F_n} with

$$
F_n = \bigcup_{k=0}^{n-1} \left[\frac{2k}{2n}, \frac{2k+1}{2n}\right].
$$

This example shows that the weak* convergence in $L^\infty[0,1]$ has nothing to do with pointwise convergence (even when we take a subsequence).

The following theorem is the most useful among results on the weak* topology.

Theorem 3.7 (Banach–Alaoglu). *The closed unit ball $B_{X^*} = \{\varphi \in X^*; \|\varphi\| \leq 1\}$ of the dual space X^* of a Banach space X is compact in the weak* topology.*

Proof. For each $x \in X$, we consider the closed disk $K_x = \{z \in \mathbb{C}; |z| \leq \|x\|\}$ and set $K = \prod_{x \in X} K_x$. Then, K is compact in the product topology by the Tychonoff theorem. We define a map $\rho : B_{X^*} \to K$ by $\rho(\varphi) = (\varphi(x))_{x \in X}$. Then, ρ is injective. It suffices to show that ρ is a homeomorphism from B_{X^*} onto $\rho(B_{X^*})$ and $\rho(B_{X^*})$ is a closed subset of K.

For $\varphi \in B_{X^*}$, $F \Subset X$, and $\varepsilon > 0$, let

$$
V(\rho(\varphi); F, \varepsilon) = \{(z_x)_{x \in X} \in K; \ \forall y \in F, \ |z_y - \varphi(y)| < \varepsilon\}.
$$

Then, $\{V(\rho(\varphi); F, \varepsilon)\}_{F \in X, \varepsilon > 0}$ is a fundamental system of neighborhoods of $\rho(\varphi)$ in K, and

$$\rho(U(\varphi; F, \varepsilon)) = V(\rho(\varepsilon); F, \varepsilon) \cap \rho(B_{X^*}).$$

Thus, ρ is a homeomorphism from B_{X^*} onto $\rho(B_{X^*})$.

For $u, v \in X$ and $\alpha \in \mathbb{C}$, let

$$L_{u,v} = \{(z_x) \in K;\ z_{u+v} = z_u + z_v\},$$

$$M_{\alpha,u} = \{(z_x) \in K;\ z_{\alpha u} = \alpha z_u\}.$$

Then, these are closed sets. From

$$\rho(B_{X^*}) = \left(\bigcap_{u,v \in X} L_{u,v}\right) \cap \left(\bigcap_{\alpha \in \mathbb{C},\ u \in L} M_{\alpha,u}\right),$$

we can see that $\rho(B_{X^*})$ is closed in K. □

Since the weak topology of a reflexive Banach space X coincides with the weak* topology of X as the dual space of X^*, we get the following.

Corollary 3.5. *The closed unit ball of a reflexive Banach space is compact in the weak topology.*

In fact, the converse is also true, which we will discuss in Chapter 5 after proving the Hahn–Banach separation theorem.

Problem 3.7. For a separable Banach space X, show that the restriction of the weak* topology to B_{X^*} is metrizable.

Exercises

Exercise 3.1
Let (Ω, μ) be a finite measure space, and let $\psi \in L^1(\Omega, \mu)^*$. Show the following:

(1) There exists $g \in L^2(\Omega, \mu)$ satisfying $\psi(f) = \int_\Omega fg\,d\mu$ for all $f \in L^2(\Omega, \mu)$.

(2) If $r > 0$ satisfies $\mu(\{|g| > r\}) \neq 0$, we have $r \leq \|\psi\|$. In particular, $g \in L^\infty(\Omega, \mu)$.
(3) $\psi(f) = \int_\Omega fg d\mu$ holds for all $f \in L^1(\Omega, \mu)$.

Exercise 3.2

Let (Ω, μ) be a σ-finite measure space, and choose a finite measure μ_0 equivalent to μ (see Exercise 2.7).

(1) For $1 \leq p < \infty$, we define $V : L^p(\Omega, \mu_0) \to L^p(\Omega, \mu)$ by $Vf = (d\mu_0/d\mu)^{1/p} f$. Show that V is an isometric isomorphism.
(2) Show $L^\infty(\Omega, \mu_0) = L^\infty(\Omega, \mu)$.
(3) Show Theorem 3.4 for $p = 1$.

Exercise 3.3

Assume that a sequence $\{x_n\}_{n=1}^\infty$ in a Banach space X weakly converges to $x \in X$. Show the following:

(1) $\|x\| \leq \liminf_{n \to \infty} \|x_n\|$.
(2) Assume, moreover, that X is a Hilbert space. Then, $\lim_{n \to \infty} \|x_n\| = \|x\|$ implies $\lim_{n \to \infty} \|x - x_n\| = 0$. (More generally, this is the case for every uniformly convex space (see Definition A.1).)

Exercise 3.4

Let Ω be a compact metric space, and let $\mathbf{P}(\Omega)$ be the set of Borel probability measures on Ω.

(1) Show that for every sequence $\{\mu_n\}_{n=1}^\infty$ in $\mathbf{P}(\Omega)$, there exists a subsequence $\{\mu_{n_k}\}_{k=1}^\infty$ and $\mu \in \mathbf{P}(\Omega)$ such that for every $f \in C(\Omega)$, the following holds:

$$\lim_{k \to \infty} \int_\Omega f d\mu_{n_k} = \int_\Omega f d\mu.$$

(2) Let $T : \Omega \to \Omega$ be a homeomorphism, and let

$$\mathbf{P}_T(\Omega) = \{\mu \in \mathbf{P}(\Omega); \ \mu \circ T = \mu\}.$$

Show that $\mathbf{P}_T(\Omega)$ is not empty.

Exercise 3.5

Show that every bounded sequence in a Hilbert space has a weakly convergent subsequence without using the Banach–Alaoglu theorem.

Exercise 3.6

In this exercise, we describe how the reflexivity of ℓ^1 and ℓ^∞ breaks down in concrete terms.

(1) For $\varphi \in \ell^{\infty*}$, define a sequence $a = (a_k)_{k=1}^\infty$ by $a_k = \varphi(\delta_k)$. Show $a \in \ell^1$ and $\varphi - \varphi_a \in c_0^\perp (= (\ell^\infty/c_0)^*)$.

(2) Show that $\ell^{\infty*}$ is isometrically isomorphic to $\ell^1 \oplus_1 (\ell^\infty/c_0)^*$.

(3) Show that $\ell^{\infty**}$ is isometrically isomorphic to $\ell^\infty \oplus_\infty (\ell^\infty/c_0)^{**}$. Here, the natural embedding from ℓ^∞ into $\ell^{\infty**}$ corresponds to the embedding into the first component of the right-hand side.

Chapter 4

Consequences of Completeness

In this chapter, we prove several results frequently used in functional analysis as applications of the Baire category theorem on complete metric spaces and explain their applications.

4.1 The Baire Category Theorem

Theorem 4.1. *Let (X, d) be a complete metric space, and let U_n, $n = 1, 2, \ldots$, be a sequence of dense open subsets of X. Then, $\bigcap_{n=1}^{\infty} U_n$ is dense in X.*

Proof. Let x be an arbitrary element in X, and let $\varepsilon > 0$ be an arbitrary positive number. Since U_1 is dense in X, there exists $x_1 \in U_1$ satisfying $d(x, x_1) < \varepsilon/2$. Since U_1 is open, there exists $0 < r_1 < \varepsilon/2$ satisfying $\overline{B(x_1, r_1)} \subset U_1$. Since U_2 is dense in X, we have $U_2 \cap B(x_1, r_1) \neq \emptyset$, which is open. Thus, there exist $x_2 \in X$ and $0 < r_2 < r_1/2$ satisfying $\overline{B(x_2, r_2)} \subset U_2 \cap B(x_1, r_1)$. Repeating the same argument, we can inductively take a sequence $\{x_n\}_{n=1}^{\infty}$ in X and a sequence of positive numbers $\{r_n\}_{n=1}^{\infty}$ such that for all $n \in \mathbb{N}$, the following hold:

- $0 < r_{n+1} < r_n/2$,
- $\overline{B(x_{n+1}, r_{n+1})} \subset U_{n+1} \cap B(x_n, r_n)$.

Then, from $r_1 < \varepsilon/2$, we get $r_n \leq 2^{-n}\varepsilon$. For $m > n$, we have $x_m \in B(x_n, r_n)$ and $d(x_m, x_n) < 2^{-n}\varepsilon$, which shows that $\{x_n\}_{n=1}^{\infty}$ is a Cauchy sequence. Since X is complete, it converges, and we denote its limit by x_∞. For $m > n$, we have $x_m \in \overline{B(x_n, r_n)}$ and $x_\infty \in \overline{B(x_n, r_n)} \subset U_n$. As this is the case for all n, we get $x_\infty \in \bigcap_{n=1}^{\infty} U_n$. Since $x_\infty \in B(x_1, r_1)$, we get $d(x_\infty, x) \leq d(x_\infty, x_1) + d(x_1, x) < \varepsilon$. $\qquad\square$

Corollary 4.1 (Baire category theorem). *Let X be a complete metric space, and let F_n, $n = 1, 2, \ldots$, be a sequence of closed subsets of X. Then, if $X = \bigcup_{n=1}^{\infty} F_n$, there exists n such that the interior of F_n is non-empty.*

Proof. We show the contraposition. Assume that F_n is a closed set without interior for all n. Then, $U_n = X \backslash F_n$ is a dense open subset of X, and $\bigcap_{n=1}^{\infty} U_n \neq \emptyset$. Thus, $\bigcup_{n=1}^{\infty} F_n \neq X$. $\qquad\square$

When a topological space is a countable union of nowhere dense (i.e., a set whose closure has no interior) subsets, we say that it is of the first category. When a topological space is not of the first category, we say that it is of the second category. A complete metric space is of the second category. This use of the term "category" is historical and has nothing to do with the category theory in modern mathematics.

4.2 Applications of the Baire Category Theorem

4.2.1 *Banach inverse mapping theorem*

Theorem 4.2 (Banach inverse mapping theorem). *Let X and Y be Banach spaces, and assume that $T \in \mathbf{B}(X, Y)$ is a bijection. Then, T^{-1} is bounded (i.e., T is an isomorphism from X onto Y).*

Proof. For $r > 0$, we set $X_r = \{x \in X \; \|x\| \leq r\}$ and $Y_r = \{y \in Y; \|y\| \leq r\}$. To prove that T^{-1} is bounded, it suffices to show that there exists $r > 0$ satisfying $T^{-1}Y_1 \subset X_r$, which is equivalent to $Y_1 \subset TX_r$. Since $\bigcup_{n=1}^{\infty} TX_n = Y$, the Baire category theorem implies that there exists $N \in \mathbb{N}$ such that the interior of $\overline{TX_N}$ is non-empty. Thus, there exist $a \in X$ and $\varepsilon > 0$ such that $Ta + Y_\varepsilon \subset \overline{TX_N}$, which is equivalent to $Y_\varepsilon \subset \overline{T(X_N - a)}$. Taking M satisfying $(N + \|a\|)/\varepsilon \leq M$, we get $Y_1 \subset \overline{TX_M}$. We show $Y_1 \subset TX_{2M}$ as follows.

Let $y_0 \in Y_1$. Since $y_0 \in \overline{TX_M}$, there exists $x_0 \in X_M$ satisfying $\|y_0 - Tx_0\| < 2^{-1}$. Letting $y_1 = y_0 - Tx_0$, we have $y_1 \in Y_{1/2} \subset \overline{TX_{M/2}}$.

In the same way, we can inductively construct a sequence $\{x_n\}_{n=0}^{\infty}$ in X such that $x_n \in X_{2^{-n}M}$ for every $n \in \mathbb{N}$ and

$$\|y_0 - T(x_0 + x_1 + \cdots + x_n)\| \leq \frac{1}{2^{n+1}}.$$

The series $x = \sum_{n=0}^{\infty} x_n$ converges and $\|x\| \leq 2M$. Since T is bounded, we get

$$Tx = \sum_{n=0}^{\infty} Tx_n = y_0.$$

Therefore, $Y_1 \subset TX_{2M}$ holds. $\qquad\square$

Example 4.1. Let M and N be closed subspaces of a Banach space X satisfying $M \cap N = \{0\}$ and $X = M + N$ (i.e., X is a direct sum of M and N as a linear space.) Then, $T : M \oplus_1 N \to X, (m, n) \mapsto m+n$, is an isomorphism of Banach spaces. Indeed, since

$$\|T(m, n)\| = \|m + n\| \leq \|m\| + \|n\| = \|(m, n)\|_1,$$

we have $T \in \mathbf{B}(M \oplus_1 N, X)$. Since T is a bijection, the Banach inverse mapping theorem implies that T^{-1} is bounded. From

$$\|m\| \leq \|(m, n)\|_1 \leq \|T^{-1}\| \|m + n\|,$$

we can also see that the projection associating m (respectively, n) with $m + n$ is bounded as well.

Problem 4.1. Let Y be a closed subspace of a Banach space X. Show that the quotient topology of X/Y coincides with the topology induced by the quotient norm. In particular, show that the quotient map $Q : X \to X/Y$ is an open mapping.

Proposition 4.1 (Open mapping theorem). *Let X and Y be Banach spaces, and assume that $T \in \mathbf{B}(X, Y)$ is a surjection. Then, T is an open mapping.*

Proof. Let $Q : X \to X/\ker T$ be the quotient map. Then, Theorem 1.5 implies that there exists $\tilde{T} \in \mathbf{B}(X/\ker T, Y)$ satisfying $T = \tilde{T} \circ Q$. Since T is a surjection, the operator $\tilde{T}$ is a bijection, and the Banach inverse mapping theorem implies that $\tilde{T}^{-1}$ is continuous. In particular, $\tilde{T}$ is an open mapping. Since Q is an open mapping too, so is T. $\qquad\square$

4.2.2 *Closed graph theorem*

Definition 4.1. Let X and Y be Banach spaces. Let $T : \mathcal{D}(T) \to Y$ be an operator from a subspace $\mathcal{D}(T)$ of X into Y. We say that $\mathcal{D}(T)$ is the *domain* of T. We define the *graph* of T by

$$\mathcal{G}(T) = \{(x, Tx) \in X \oplus_1 Y; \ x \in \mathcal{D}(T)\}.$$

We say that T is a *closed operator* if $\mathcal{G}(T)$ is a closed subspace of $X \oplus_1 Y$.

We can rephrase the condition that T is a closed operator in terms of sequences as follows: Whenever a sequence $\{x_n\}_{n=1}^{\infty}$ in $\mathcal{D}(T)$ converges to $x \in X$ and $\{Tx_n\}_{n=1}^{\infty}$ converges to $y \in Y$, we have $x \in \mathcal{D}(T)$ and $Tx = y$.

While we used $X \oplus_1 Y$ as a Banach space direct sum of X and Y in the above definition, we can get an equivalent definition if we replace $X \oplus_1 Y$ with $X \oplus_p Y$ for $1 \le p \le \infty$. In fact, it is more natural to use $X \oplus_2 Y$ when X and Y are Hilbert spaces.

Example 4.2. Let $X = Y = C[0,1]$, and define an operator $T : \mathcal{D}(T) \to C[0,1]$ with $\mathcal{D}(T) = C^1[0,1]$ by $Tf(t) = f'(t)$. Then, T is a closed operator. Indeed, assume that $\{f_n\}_{n=1}^{\infty}$ is a sequence of functions in $C^1[0,1]$ such that $\{f_n\}_{n=1}^{\infty}$ uniformly converges to $f \in C[0,1]$ and $\{f_n'\}_{n=1}^{\infty}$ uniformly converges to $g \in C[0,1]$. Since

$$f_n(t) = f_n(0) + \int_0^t f_n'(s)ds,$$

we get

$$f(t) = f(0) + \int_0^t g(s)ds.$$

Thus, $f \in \mathcal{D}(T)$ and $Tf = g$.

As this example suggests, a typical example of a closed operator is unbounded. However, the situation is different when the domain is the whole space X.

Theorem 4.3 (Closed graph theorem). *Let X and Y be Banach spaces, and let $T : X \to Y$ be a closed operator whose domain is the whole X. Then, T is bounded.*

Proof. Note that the graph $\mathcal{G}(T)$ of T is a Banach space, as it is a closed subspace of $X \oplus_1 Y$. Let $P : \mathcal{G}(T) \to X$ be the projection onto the first component. Then, P is a bounded bijection, and the Banach inverse mapping theorem implies that P^{-1} is bounded as well. Since $P^{-1}x = (x, Tx)$, we get

$$\|Tx\| \leq \|x\| + \|Tx\| = \|(x, Tx)\|_1 \leq \|P^{-1}\|\|x\|,$$

and T is bounded. $\qquad\qquad\qquad\qquad\qquad\qquad\qquad\qquad\qquad\square$

Example 4.3. Let (Ω, μ) be a σ-finite measure space, and let $f : \Omega \to \mathbb{C}$ be a measurable function satisfying $fg \in L^2(\Omega, \mu)$ for all $g \in L^2(\Omega, \mu)$. Then, $f \in L^\infty(\Omega, \mu)$. While this fact may be shown by a purely measure-theoretic argument, here we give a simple argument using the closed graph theorem.

We introduce an operator $T : L^2(\Omega, \mu) \to L^2(\Omega, \mu)$ by $Tg = fg$, and we show that T is a closed operator first. We assume that a sequence of functions $\{g_n\}_{n=1}^\infty$ in $L^2(\Omega, \mu)$ and $g, h \in L^2(\Omega, \mu)$ satisfy $\lim_{n\to\infty} \|g_n - g\|_2 = 0$ and $\lim_{n\to\infty} \|Tg_n - h\|_2 = 0$. Then, since there exists a subsequence $\{g_{n_k}\}_{k=1}^\infty$ of $\{g_n\}_{n=1}^\infty$ such that $\{g_{n_k}\}_{k=1}^\infty$ converges to g almost everywhere and $\{fg_{n_k}\}_{k=1}^\infty$ converges to h almost everywhere (see Corollary A.1), we get $h = fg$. This means that T is a closed operator, and as $\mathcal{D}(T) = L^2(\Omega, \mu)$, the closed graph theorem implies that T is bounded. We assume $f \notin L^\infty(\Omega, \mu)$ and deduce a contradiction as follows. From the assumption, the measure of $E = \{|f| > \|T\| + 1\}$ is not 0. Since (Ω, ν) is σ-finite, there exists a measurable subset E_0 of E satisfying $0 < \mu(E_0) < \infty$. Let $g = \mu(E_0)^{-1/2}\chi_{E_0}$. Then, $\|g\|_2 = 1$ and

$$\|Tg\|_2^2 = \frac{1}{\mu(E_0)} \int_{E_0} |f|^2 d\mu \geq \frac{1}{\mu(E_0)} \int_{E_0} (\|T\| + 1)^2 d\mu = (\|T\| + 1)^2,$$

which is a contradiction. Therefore, $f \in L^\infty(\Omega, \mu)$.

4.2.3 *The principle of uniform boundedness*

Before getting into the main topic, we note that for a set Λ and a Banach space Y, if we set

$$\ell^\infty(\Lambda, Y) = \left\{ (y_\lambda)_{\lambda \in \Lambda} \in Y^\Lambda; \ \sup_{\lambda \in \Lambda} \|y_\lambda\| < \infty \right\},$$

then $\ell^\infty(\Lambda, Y)$ is a Banach space with pointwise operations and the norm $\|(y_\lambda)_{\lambda \in \Lambda}\|_\infty = \sup_{\lambda \in \Lambda} \|y_\lambda\|$. The proof is not so different from the case of ℓ^∞.

Theorem 4.4 (The principle of uniform boundedness). *Let X and Y be Banach spaces, and assume that a family of bounded operators $\{T_\lambda\}_{\lambda \in \Lambda} \subset \mathbf{B}(X, Y)$ satisfies $\sup_{\lambda \in \Lambda} \|T_\lambda x\| < \infty$ for every $x \in X$. Then, $\{\|T_\lambda\|\}_{\lambda \in \Lambda}$ is bounded.*

Proof. We define an operator $T : X \to \ell^\infty(\Lambda, Y)$ by $Tx = (T_\lambda x)_{\lambda \in \Lambda}$ and show that T is a closed operator. Assume that $\{x_n\}_{n=1}^\infty \subset X$, $x \in X$, and $y = (y_\lambda)_\Lambda \in \ell^\infty(\Lambda, Y)$ satisfy $\lim_{n \to \infty} \|x_n - x\| = 0$ and $\lim_{n \to \infty} \|Tx_n - y\| = 0$. Since

$$\lim_{n \to \infty} \sup_{\lambda \in \Lambda} \|T_\lambda x_n - y_\lambda\| = 0,$$

we have, in particular, $\lim_{n \to \infty} \|T_\lambda x_n - y_\lambda\|$ for each $\lambda \in \Lambda$, and so $T_\lambda x = y_\lambda$. Thus, $Tx = y$, and T is a closed operator. Since $\mathcal{D}(T) = X$, the closed graph theorem implies that T is bounded. Thus, $\|T_\lambda x\| \leq \|T\|\|x\|$ holds for all $x \in X$ and $\lambda \in \Lambda$, and $\|T_\lambda\| \leq \|T\|$. $\square$

Corollary 4.2 (Banach–Steinhaus). *Let X and Y be Banach spaces, let $\{T_n\}_{n=1}^\infty$ be a sequence in $\mathbf{B}(X, Y)$, and assume that $\{T_n x\}_{n=1}^\infty$ converges for all $x \in X$. Then, $\{\|T_n\|\}_{n=1}^\infty$ is bounded, and if we define T by $Tx = \lim_{n \to \infty} T_n x$, then $T \in \mathbf{B}(X, Y)$.*

Proof. As $\{T_n x\}_{n=1}^\infty$ converges, the sequence $\{\|T_n x\|\}_{n=1}^\infty$ is bounded, and the statement follows from the previous theorem. $\square$

When $Y = \mathbb{C}$, the above statement takes the following form. Let $\{\varphi_n\}_{n=1}^\infty$ be a sequence in X^* and assume that $\{\varphi_n(x)\}_{n=1}^\infty$ converges for every $x \in X$. Then, $\{\|\varphi_n\|\}_{n=1}^\infty$ is bounded and $\{\varphi_n\}_{n=1}^\infty$ converges to some $\varphi \in X^*$ in the weak* topology. It also follows that every convergent sequence in the weak* topology is bounded.

Corollary 4.3. *Let $\{x_n\}_{n=1}^\infty$ be a sequence in a Banach space X and assume that $\{\varphi(x_n)\}_{n=1}^\infty$ converges for every $\varphi \in X^*$. Then, $\{x_n\}_{n=1}^\infty$ is bounded. In particular, every weakly convergent sequence is bounded.*

Proof. The statement follows from the previous corollary if we regard $\{x_n\}_{n=1}^\infty$ as a sequence in $(X^*)^*$ by $X \subset X^{**}$. $\square$

Definition 4.2. Let $\mathcal{H}$ be a Hilbert space, let $\{T_n\}_{n=1}^\infty$ be a sequence in $\mathbf{B}(H)$, and let $T \in \mathbf{B}(\mathcal{H})$.

- If $\{T_n x\}_{n=1}^\infty$ converges to Tx for all $x \in \mathcal{H}$, we say that $\{T_n\}_{n=1}^\infty$ *strongly converges* to T and write s-$\lim_{n\to\infty} T_n = T$.
- If $\{\langle T_n x, y \rangle\}_{n=1}^\infty$ converges to $\langle Tx, y \rangle$ for all $x, y \in \mathcal{H}$, we say that $\{T_n\}_{n=1}^\infty$ *weakly converges* to T and write w-$\lim_{n\to\infty} T_n = T$. This is equivalent to the condition that for every $x \in \mathcal{H}$, the sequence $\{T_n x\}_{n=1}^\infty$ weakly converges to Tx (in the sense of $\mathcal{H}$).

Remark 4.1. A word of caution: The above definition of the weak convergence of a sequence of operators is *different* from the convergence in the weak topology of the Banach space $\mathbf{B}(\mathcal{H})$. However, weak convergence in the latter sense is rarely used and it seldom causes confusion.

In what follows, we call convergence in $\mathbf{B}(\mathcal{H})$ in the operator norm *norm convergence*. From the definitions, the following holds for the convergences of sequences of operators:

norm convergence $\implies$ strong convergence $\implies$ weak convergence.

The converse implication does not hold in either case.

Example 4.4. Let V be the unilateral shift of ℓ^2.

- The sequence $\{V^{*n}\}_{n=1}^\infty$ strongly converges to 0 but does not converge in norm. In fact, since $(V^{*n}x)_k = x_{n+k}$, we have

$$\lim_{n\to\infty} \|V^{*n}x\|^2 = \lim_{n\to\infty} \sum_{k=n+1}^\infty |x_k|^2 = 0.$$

On the other hand, since V^n is an isometry, we have $\|V^n\| = 1$ and $\|V^{*n}\| = \|V^n\| = 1$.

- The sequence $\{V^n\}_{n=1}^\infty$ weakly converges to 0 but does not strongly converge. In fact, for $x, y \in \ell^2$, we have

$$|\langle V^n x, y \rangle| = |\langle x, V^{*n}y \rangle| \le \|x\|\|V^{*n}y\|,$$

and $\{V^n\}_{n=1}^\infty$ weakly converges to 0. Since V^n is an isometry, we have $\|V^n x\| = \|x\|$.

Theorem 4.5. *Let $\mathcal{H}$ be a Hilbert space. If a sequence $\{T_n\}_{n=1}^{\infty}$ in $\mathbf{B}(\mathcal{H})$ weakly converges to $T \in \mathbf{B}(\mathcal{H})$, then $\{\|T_n\|\}_{n=1}^{\infty}$ is bounded.*

Proof. For each $x \in \mathcal{H}$, the sequence $\{T_n x\}_{n=1}^{\infty}$ weakly converges to Tx, and the principle of uniform boundedness implies that $\{\|T_n x\|\}_{n=1}^{\infty}$ is bounded. Thus, the principle of uniform boundedness again implies that $\{\|T_n\|\}_{n=1}^{\infty}$ is bounded. $\qquad\square$

Proposition 4.2. *Let $\mathcal{H}$ be a Hilbert space, and let $\{S_n\}_{n=1}^{\infty}$ and $\{T_n\}_{n=1}^{\infty}$ be sequences in $\mathbf{B}(\mathcal{H})$ strongly converging to $S \in \mathbf{B}(\mathcal{H})$ and $T \in \mathbf{B}(\mathcal{H})$, respectively. Then, $\{S_n T_n\}_{n=1}^{\infty}$ strongly converges to ST.*

Proof. By the principle of uniform boundedness, there exists $M > 0$ satisfying $\|S_n\| \leq M$ for all $n \in \mathbb{N}$. Thus, for every $x \in \mathcal{H}$, we have

$$
\begin{aligned}
\|S_n T_n x - ST x\| &= \|S_n(T_n x - Tx) + S_n(Tx) - S(Tx)\| \\
&\leq M\|T_n x - Tx\| + \|S_n(Tx) - S(Tx)\| \\
&\to 0 \quad (n \to \infty),
\end{aligned}
$$

which shows that $\{S_n T_n\}_{n=1}^{\infty}$ strongly converges to ST. $\qquad\square$

We end this chapter by showing the following fact as an application of the principle of uniform boundedness to the theory of Fourier series: There exists a continuous function of period 2π whose Fourier series does not converge pointwise.

In the following argument, we use the notation in Example 2.8. For $f \in C(\mathbb{T})$ and $N \in \mathbb{N}$, let

$$
S_N(f)(t) = \sum_{n=-N}^{N} \hat{f}(n) e^{int}.
$$

The problem is whether $\{S_N(f)(t)\}_{N=1}^{\infty}$ converges to $f(t)$ pointwise. Let $\varphi_N(f) = S_N(f)(0)$. It suffices to show that there exists $f \in C(\mathbb{T})$ such that $\{\varphi_N(f)\}_{N=1}^{\infty}$ does not converge.

Now, we have

$$
\varphi_N(f) = \sum_{n=-N}^{N} \frac{1}{2\pi} \int_{-\pi}^{\pi} f(t) e^{-int} dt = \frac{1}{2\pi} \int_{-\pi}^{\pi} f(t) D_N(t) dt.
$$

Here, $D_N(t)$ is called the *Dirichlet kernel*

$$D_N(t) = \frac{\sin(N + \frac{1}{2})t}{\sin \frac{t}{2}}.$$

From this formula, we see that $\varphi_N \in C(\mathbb{T})^*$ and

$$\|\varphi_N\| \leq \frac{1}{2\pi} \int_{-\pi}^{\pi} |D_N(t)| dt = \|D_N\|_1.$$

Problem 4.2. In the above situation, show the following:

(1) $\|\varphi_N\| = \|D_N\|_1$.
(2) $\{\|D_N\|_1\}_{N=1}^{\infty}$ is not bounded.
(3) There exists $f \in C(\mathbb{T})$ such that $\{\varphi_N(f)\}_{N=1}^{\infty}$ does not converge.

Exercises

Exercise 4.1
A countable intersection of open subsets is called a G_δ-*set*. Show that in a complete metric space, a countable intersection of dense G_δ-sets is again a dense G_δ-set.

Exercise 4.2
Let $\mathcal{H}$ be a Hilbert space, and assume that an operator $A : \mathcal{H} \to \mathcal{H}$ satisfies $\langle Ax, y \rangle = \langle x, Ay \rangle$ for all $x, y \in \mathcal{H}$. Show $A \in \mathbf{B}(\mathcal{H})_{\mathrm{sa}}$.

Exercise 4.3
Let X be a Banach space, and let $D \subset \mathbb{C}$ be a domain. Assume that a map $f : D \to X$ satisfies the following condition: For every $\varphi \in X^*$, the function $\varphi(f(z))$ is holomorphic on D.

(1) Show that for every $\zeta \in D$ and every sequence of complex numbers $\{h_n\}_{n=1}^{\infty}$ converging to 0 and satisfying $h_n \neq 0$ for all $n \in \mathbb{N}$, the sequence

$$\left\{ \frac{1}{h_n}(f(\zeta + h_n) - f(\zeta)) \right\}_{n=1}^{\infty}$$

converges to some element in X^{**} in the weak* topology.

(2) Show that there exists a map $g : D \to X^{**}$ such that

$$\frac{d\varphi(f(z))}{dz} = \varphi(g(z))$$

holds for all $\varphi \in X^*$.

(3) Show $\lim_{n\to\infty} h_n^{-1}(f(\zeta + h_n) - f(\zeta)) = g(\zeta)$. In particular, show $g(\zeta) \in X$.

Exercise 4.4

Let U be the bilateral shift of $\ell^2(\mathbb{Z})$. Show that $\{U^n\}_{n=1}^\infty$ weakly converges to 0.

Exercise 4.5

Let $\mathcal{H}$ be a Hilbert space. For $T \in \mathbf{B}(\mathcal{H})$ and $n \in \mathbb{N}$, let $A_n(T) = n^{-1}\sum_{k=0}^{n-1} T^k$.

(1) If $\{\|T^n\|\}_{n=1}^\infty$ is bounded and $\mathcal{R}(I - T)$ is dense in $\mathcal{H}$ (e.g., the unilateral shift of ℓ^2), show that $\{A_n(T)\}_{n=1}^\infty$ strongly converges to 0.

(2) Show that for every $U \in \mathcal{U}(\mathcal{H})$, the sequence $\{A_n(U)\}_{n=1}^\infty$ strongly converges to the projection onto $\ker(I - U)$ (von Neumann's mean ergodic theorem).

Chapter 5

Locally Convex Spaces#

Even if we decide to restrict ourselves to considering only Banach spaces in functional analysis, we still need locally convex spaces for considering the weak and weak* topologies. In this chapter, after we explain the basics of locally convex spaces, we show the Hahn–Banach separation theorem and illustrate a few applications of it.

5.1 Definition of Locally Convex Spaces and Basic Examples

Definition 5.1. We say that X is a *topological linear space* if X is a linear space and a topological space satisfying the following conditions:

(1) The map $X \times X \ni (x, y) \mapsto x + y \in X$ is continuous.
(2) The map $\mathbb{C} \times X \ni (\alpha, x) \mapsto \alpha x \in X$ is continuous.

If X is a topological linear space:

- for a fixed $\alpha \in \mathbb{C} \backslash \{0\}$, the map $X \ni x \mapsto \alpha x \in X$ is a homeomorphism;
- for a fixed $a \in X$, the map $X \ni x \mapsto x + a \in X$ is a homeomorphism.

If X and Y are topological linear spaces and $T : X \to Y$ is a linear map, from the continuity of translation, we see that T is continuous if and only if T is continuous at one point.

The following lemma holds for a general topological group.

Lemma 5.1. *For a topological linear space X, the following conditions are equivalent:*

(1) X *is a Hausdorff space.*
(2) *For every $x \in X \backslash \{0\}$, there exists a neighborhood U of 0 satisfying $x \notin U$.*

Proof. Since the implication (1) $\Longrightarrow$ (2) is trivial, we show only (2) $\Longrightarrow$ (1). Let $a, b \in X$, $a \neq b$. Since $b - a \neq 0$, there exists a neighborhood U of 0 satisfying $b - a \notin U$. Let $U_1 = a - b + U$. Then, U_1 is a neighborhood of $a - b$ and $0 \notin U_1$. Since the map $X \times X \ni (x, y) \mapsto x - y \in X$ is continuous, and in particular continuous at (a, b), there exist a neighborhood V of a and a neighborhood W of b satisfying $V - W \subset U_1$. Since $0 \notin U_1$, we have $V \cap W = \emptyset$. $\qquad \square$

In what follows, the topological linear spaces we consider are assumed to be Hausdorff.

Definition 5.2. We say that a topological linear space X is a *locally convex space* if X has a fundamental system of neighborhoods consisting of convex sets.

Example 5.1. Let X be a normed space.

(1) With the topology given by the norm (called the norm topology in what follows), X is a locally convex space, and $\{B(0, \varepsilon)\}_{\varepsilon > 0}$ is a fundamental system of neighborhoods of 0 consisting of convex sets.
(2) With the weak topology, X is a locally convex space, and $\{U(0; F, \varepsilon)\}_{F \Subset X^*, \varepsilon > 0}$ is a fundamental system of neighborhoods of 0 consisting of convex sets.
(3) With the weak* topology, X^* is a locally convex space, and $\{U(0; F, \varepsilon)\}_{F \Subset X, \varepsilon > 0}$ is a fundamental system of neighborhoods of 0 consisting of convex sets.

Let X be a locally convex space, and let U be a convex neighborhood of 0. Since for a fixed $x \in X$, the map $\mathbb{C} \to X$, $\alpha \mapsto \alpha x$, is continuous and $0 \cdot x = 0$, there exists $\delta > 0$ such that $|\alpha| < \delta$ implies

$\alpha x \in U$. Thus, we can define a function $p_U : X \to [0, \infty)$ by

$$p_U(x) = \inf \left\{ t > 0; \ \frac{1}{t} x \in U \right\}.$$

We call this function p_U the *Minkowski functional* of U. Since U is a convex set containing 0, for every $t > p_U(x)$, we have $t^{-1}x \in U$.

Lemma 5.2. *For the Minkowski functional p_U, the following hold:*

(1) $p_U(x + y) \leq p_U(x) + p_U(y)$ *holds for all $x, y \in X$.*
(2) $p_U(tx) = tp_U(x)$ *holds for all $x \in X$ and $t > 0$.*
(3) *We have $\{x \in X; \ p_U(x) < 1\} \subset U \subset \{x \in X; \ p_U(x) \leq 1\}$.*

Proof. Since (2) and (3) directly follow from the definition, we show only (1). We take $\varepsilon > 0$ and set $x' = (p_U(x) + \varepsilon)^{-1}x$ and $y' = (p_U(y) + \varepsilon)^{-1}y$. Then, $x', y' \in U$. Since U is a convex set, we have

$$\frac{1}{p_U(x) + p_U(y) + 2\varepsilon}(x + y)$$

$$= \frac{p_U(x) + \varepsilon}{p_U(x) + p_U(y) + 2\varepsilon}x' + \frac{p_U(y) + \varepsilon}{p_U(x) + p_U(y) + 2\varepsilon}y' \in U,$$

and $p_U(x + y) \leq p_U(x) + p_U(y) + 2\varepsilon$ holds. As $\varepsilon > 0$ is arbitrary, we get (1). □

Remark 5.1. Note that (1) and (2) are the conditions that appear in the Hahn–Banach extension theorem (in the real case). (3) Shows that U can be more or less recovered from p_U.

Let X be a topological linear space. We say that a convex neighborhood U of $0 \in X$ is *absolutely convex* if $\alpha U = U$ holds for all complex numbers α with modulus 1. If U is absolutely convex, its Minkowski functional p_U is a semi-norm.

Definition 5.3. Let X be a linear space, and let $\{p_\lambda\}_{\lambda \in \Lambda}$ be a family of semi-norms of X satisfying the following condition: $\forall x \in X \setminus \{0\}$, $\exists \lambda_0 \in \Lambda$, $p_{\lambda_0}(x) \neq 0$. Then, for $x \in X$, a finite subset $F \Subset \Lambda$, and $\varepsilon > 0$, we set

$$U(x; F, \varepsilon) = \{y \in X; \ \forall \lambda \in F, \ p_\lambda(y - x) < \varepsilon\},$$

and we introduce a topology into X for which $\{U(x; F, \varepsilon)\}_{F \in \Lambda,\ \varepsilon > 0}$ is a fundamental system of neighborhoods of x. We call this topology the *topology given by the family of semi-norms* $\{p_\lambda\}_{\lambda \in \Lambda}$.

Remark 5.2. The above $U(0; F, \varepsilon)$ is an absolutely convex set. Also, if $x \in X \backslash \{0\}$ and $p_{\lambda_0}(x) \neq 0$, we have $x \notin U(0; \{\lambda_0\}, p_{\lambda_0}(x))$. Thus, Lemma 5.1 shows that the locally convex topology of X given by $\{p_\lambda\}_{\lambda \in \Lambda}$ is Hausdorff.

Summing up the facts we have seen above, we get the following theorem.

Theorem 5.1. *For a topological linear space X, the following conditions are equivalent*:

(1) *The topology of X is given by a family of semi-norms.*
(2) *There exists a fundamental system of neighborhoods of 0 consisting of absolutely convex neighborhoods.*

Problem 5.1. Show the above theorem.

Example 5.2. Let X be a normed space.

(1) The norm topology of X is the locally convex topology given by one semi-norm $p(x) = \|x\|$.
(2) For $\varphi \in X^*$, we define a semi-norm of X by $p_\varphi(x) = |\varphi(x)|$. Then, the weak topology of X is the locally convex topology given by $\{p_\varphi\}_{\varphi \in X^*}$.
(3) For $x \in X$, we define a semi-norm of X^* by $p_x(\varphi) = |\varphi(x)|$. Then, the weak* topology of X^* is the locally convex topology given by $\{p_x\}_{x \in X}$.

Example 5.3. For $f \in C^\infty(\mathbb{R})$ and $n \in \mathbb{N}_0$, we define

$$p_n(f) = \max_{0 \leq k, l \leq n} \sup_{x \in \mathbb{R}} |x|^k |f^{(l)}(x)| \in [0, \infty].$$

We denote by $\mathcal{S}(\mathbb{R})$ the set of $f \in C^\infty(\mathbb{R})$ with finite $p_n(f)$ for all n and call it the **Schwartz space**. It is known that $\mathcal{S}(\mathbb{R})$ is complete with respect to the metric

$$d(f,g) = \sum_{n=0}^{\infty} \frac{1}{2^n} \frac{p_n(f-g)}{1+p_n(f-g)},$$

and the topology given by this metric coincides with the locally convex topology given by the family of semi-norms $\{p_n\}_{n=0}^{\infty}$. Such a space is called a *Fréchet space*. A continuous linear functional on $\mathcal{S}(\mathbb{R})$ is called a *tempered distribution*.

Example 5.4. Let $\mathcal{H}$ be a Hilbert space. For $x \in \mathcal{H}$, we define a semi-norm of $\mathbf{B}(\mathcal{H})$ by $p_x(T) = \|Tx\|$. The locally convex topology of $\mathbf{B}(\mathcal{H})$ given by $\{p_x\}_{x\in\mathcal{H}}$ is called the *strong operator topology*. The strong convergence of a sequence of operators introduced in Definition 4.2 is convergence in the strong operator topology.

For $x, y \in \mathcal{H}$, we define a semi-norm of $\mathbf{B}(\mathcal{H})$ by $p_{x,y}(T) = |\langle Tx, y \rangle|$. The locally convex topology of $\mathbf{B}(\mathcal{H})$ given by $\{p_{x,y}\}_{x,y\in\mathcal{H}}$ is called the *weak operator topology*. The weak convergence of a sequence of operators introduced in Definition 4.2 is convergence in the weak operator topology.

Remark 5.3. One should not rush to conclude from Proposition 4.2 that the product $\mathbf{B}(\mathcal{H}) \times \mathbf{B}(\mathcal{H}) \to \mathbf{B}(\mathcal{H})$, $(S, T) \mapsto ST$, is continuous in the strong operator topology. In fact, it is known not to be continuous (Kadison and Ringrose, 1997, Exercise 2.8.33). Consider the reason why these two facts are consistent.

5.2 Continuous Linear Functionals

We have seen in Chapter 1 that a linear functional on a normed space is continuous if and only if it is bounded. This fact is generalized in the following way.

Theorem 5.2. *When the topology of a locally convex space X is given by a family of semi-norms, $\{p_\lambda\}_{\lambda\in\Lambda}$, for a linear functional φ on X, the following conditions are equivalent:*

(1) *φ is continuous.*
(2) *φ is continuous at 0.*

(3) $\exists \lambda_1, \lambda_2, \ldots, \lambda_n \in \Lambda, \ \exists c_1, c_2, \ldots, c_n > 0, \ \forall x \in X,$

$$|\varphi(x)| \leq \sum_{i=1}^{n} c_i p_{\lambda_i}(x).$$

Proof. (1) $\Longrightarrow$ (2) is trivial.

(2) $\Longrightarrow$ (3). If φ is continuous at 0, there exists a neighborhood $U(0; F, \varepsilon)$ of 0 satisfying $|\varphi(y)| < 1$ for every $y \in U(0; F, \varepsilon)$. Here, $\varepsilon > 0$, and $F \Subset \Lambda$ is a finite subset, say $F = \{\lambda_1, \lambda_2, \ldots, \lambda_n\}$. Let

$$c_1 = c_2 = \cdots = c_n = \frac{2}{\varepsilon}.$$

We show that (3) holds for $x \in X$.

If $1 \leq \forall i \leq n, \ p_{\lambda_i}(x) = 0$, we have $tx \in U(0; F, \varepsilon)$ for every $t > 0$, and $|\varphi(tx)| < 1$, and so $\varphi(x) = 0$. Thus, (3) holds. If $1 \leq \exists i \leq n$, $p_{\lambda_i}(x) \neq 0$, we have

$$\frac{\varepsilon}{2} \frac{1}{p_{\lambda_1}(x) + p_{\lambda_2}(x) + \cdots + p_{\lambda_n}(x)} x \in U(0; F, \varepsilon)$$

and

$$\left| \varphi \left(\frac{\varepsilon}{2} \frac{1}{p_{\lambda_1}(x) + p_{\lambda_2}(x) + \cdots + p_{\lambda_n}(x)} x \right) \right| < 1,$$

and so (3) holds too.

(3) $\Longrightarrow$ (1). Let $\varepsilon > 0$. We set $F = \{\lambda_1, \lambda_2, \ldots, \lambda_n\}$ and

$$\delta = \frac{\varepsilon}{2} \frac{1}{c_1 + c_2 + \cdots + c_n}.$$

Then, $|\varphi(x)| < \varepsilon$ holds for every $x \in U(0; F, \delta)$ and φ is continuous at 0. $\qquad\square$

The following lemma is useful for applying the above theorem to concrete examples of spaces.

Lemma 5.3. *Let X be a linear space, and let $\varphi_1, \varphi_2, \ldots, \varphi_n, \psi$ be linear functionals on X. Then, $\bigcap_{i=1}^{n} \ker \varphi_i \subset \ker \psi$ implies that ψ is a linear combination of $\{\varphi_i\}_{i=1}^{n}$.*

Proof. If we choose $1 \leq i_1 < i_2 < \cdots < i_m \leq n$ so that $\{\varphi_{i_j}\}_{j=1}^m$ is a basis for $\mathrm{span}\{\varphi_i\}_{i=1}^n$, we get

$$\bigcap_{j=1}^m \ker \varphi_{i_j} = \bigcap_{i=1}^n \ker \varphi_i \subset \ker \psi.$$

Thus, we may assume that $\{\varphi_i\}_{i=1}^n$ is linearly independent from the beginning.

We first show that there exist $x_1, x_2, \ldots, x_n \in X$ satisfying $\varphi_i(x_j) = \delta_{i,j}$. Defining a linear map $T : X \to \mathbb{C}^n$ by $Tx = (\varphi_1(x), \varphi_2(x), \ldots, \varphi_n(x))$, it suffices to show that T is a surjection. Since TX is a linear subspace of $\mathbb{C}^n$, if $TX \neq \mathbb{C}^n$, there would exist $(a_1, a_2, \ldots, a_n) \in \mathbb{C}^n \setminus \{0\}$ satisfying

$$\sum_{i=1}^n a_i \varphi_i(x) = 0$$

for all $x \in X$. This contradicts the assumption that $\{\varphi_i\}_{i=1}^n$ is linearly independent, and so T is surjective.

Since we have

$$x - \sum_{j=1}^n \varphi_j(x) x_j \in \bigcap_{i=1}^n \ker \varphi_i \subset \ker \psi,$$

for every $x \in X$, we get

$$\psi(x) = \sum_{j=1}^n \psi(x_j) \varphi_j(x),$$

which shows $\psi \in \mathrm{span}\{\varphi_i\}_{i=1}^n$. $\qquad\square$

Theorem 5.3. *Let X and Y be linear spaces, and let $b : X \times Y \to \mathbb{C}$ be a non-degenerate bilinear form. For a linear functional φ on X, the following conditions are equivalent:*

(1) φ is continuous with respect to the topology $\sigma(X, Y)$.
(2) $\exists y \in Y, \forall x \in X, \varphi(x) = b(x, y)$.

Proof. For $y \in Y$, we define a semi-norm of X by $p_y(x) = |b(x, y)|$. Note that $\sigma(X, Y)$ is the locally convex topology given by the family of semi-norms $\{p_y\}_{y \in Y}$.

(1) $\Longrightarrow$ (2). By Theorem 5.2, there exist $y_1, y_2, \ldots, y_n \in Y$ and $c_1, c_2, \ldots, c_n > 0$ such that for every $x \in X$,

$$|\varphi(x)| \le c_1 p_{y_1}(x) + c_2 p_{y_2}(x) + \cdots + c_n p_{y_n}(x).$$

Setting $\varphi_i(x) = b(x, y_i)$, we get $\bigcap_{i=1}^n \ker \varphi_i \subset \ker \varphi$, and φ is a linear combination of $\{\varphi_i\}_{i=1}^n$. Thus, (2) holds.

(2) $\Longrightarrow$ (1). Since $|\varphi(x)| = p_y(x)$, Theorem 5.2 implies that φ is continuous. $\qquad\square$

Corollary 5.1. *Let X be a normed space.*

(1) *For a linear functional φ on X, the following conditions are equivalent:*

  (i) $\varphi \in X^*$.
  (ii) φ *is continuous in the weak topology.*

(2) *For a linear functional Φ on X^*, the following conditions are equivalent:*

  (i) $\Phi \in X$.
  (ii) Φ *is continuous in the weak* topology.*

5.3 Hahn–Banach Separation Theorem and Its Applications

5.3.1 *Hahn–Banach separation theorem*

Theorem 5.4 (Hahn–Banach separation theorem). *Let X be a locally convex space, and let C be a closed convex subset of X. For every $a \in X \backslash C$, there exists a continuous linear functional $\varphi : X \to \mathbb{C}$ satisfying*

$$\operatorname{Re} \varphi(a) \lneqq \inf \{\operatorname{Re} \varphi(x); \ x \in C\}.$$

Proof. Considering $(C - a, 0)$ instead of (C, a), we may and do assume $a = 0$ in the proof. We have $0 \notin C$ in this case.

Since C is closed, there exists a convex neighborhood U of 0 satisfying $U \cap C = \emptyset$. We fix $c_0 \in C$ and set

$$V = c_0 + \bigcup_{\lambda \geq 1} \lambda(U - C),$$

where $U - C = \{u - c; \ u \in U, \ c \in C\}$. Since $V \supset c_0 + U - C \supset U$, the set V is a neighborhood of 0. We show that V is a convex set. Indeed, since $C - U$ is a convex set, we get the following for $\lambda, \mu \geq 1$, $x, y \in C - U$, and $0 < t < 1$:

$$t(c_0 + \lambda x) + (1 - t)(c_0 + \mu y) = c_0 + t\lambda x + (1 - t)\mu y$$

$$= c_0 + (t\lambda + (1 - t)\mu) \left(\frac{t\lambda}{t\lambda + (1 - t)\mu} x + \frac{(1 - t)\mu}{t\lambda + (1 - t)\mu} y \right) \in V.$$

Let $p_V : X \to [0, \infty)$ be the Minkowski functional of V. Since $U \cap C = \emptyset$, we have $0 \notin U - C$, and so $c_0 \notin V$, which shows $p_V(c_0) \geq 1$. We define a real linear functional ψ_0 on the one-dimensional real subspace $\mathbb{R}c_0$ of X by $\psi_0(tc_0) = tp_V(c_0)$. Then, we have $\psi_0(x_0) \leq p_V(x_0)$ for every $x_0 \in \mathbb{R}c_0$. Thus, the Hahn–Banach extension theorem implies that there exists a real linear functional ψ on X such that it is an extension of ψ_0 and $\psi(x) \leq p_V(x)$ holds for every $x \in X$. We claim that ψ is continuous. Indeed, for a given $\varepsilon > 0$, we set $W = (\varepsilon V) \cap (-\varepsilon V)$, which is a neighborhood of 0. For every $x \in W$, we have $\psi(x) \leq p_V(x) \leq \varepsilon$ and $\psi(-x) \leq p_V(-x) \leq \varepsilon$. This shows $\psi(x) \in (-\varepsilon, \varepsilon)$ and ψ is continuous. Letting $\varphi(x) = \psi(x) - i\psi(ix)$, we get a continuous complex linear functional φ of X satisfying $\operatorname{Re} \varphi(x) = \psi(x)$.

For $u \in U$, $c \in C$, and $\lambda \geq 1$, we have

$$\psi(c_0 + \lambda(u - c)) \leq p_V(c_0 + \lambda(u - c)) \leq 1,$$

and $\lambda(\psi(u) - \psi(c)) \leq 1 - \psi(c_0)$. Since this holds for every $\lambda \geq 1$, we get $\psi(u) - \psi(c) \leq 0$. Since ψ is a continuous real linear functional with $\psi \neq 0$ and U is a neighborhood of 0, there exists $u_0 \in U$ satisfying $0 < \psi(u_0)$. Thus, we get $0 < \psi(u_0) \leq \inf\{\psi(c); \ c \in C\}$. $\qquad\square$

Remark 5.4. When C is a closed subspace in the above theorem, as $\operatorname{Re} \varphi(C)$ is bounded below, it is $\{0\}$ and φ restricted to C is 0.

Corollary 5.2. *For a convex subset C of a normed space X, the following conditions are equivalent:*

(1) C *is closed in the weak topology.*
(2) C *is closed in the norm topology.*

Proof. Comparison of the two topologies shows that the implication (1) $\Longrightarrow$ (2) is trivial, and we show only (2) $\Longrightarrow$ (1). For $a \in X \backslash C$, the Hahn–Banach separation theorem implies that there exists $\varphi \in X^*$ satisfying

$$\operatorname{Re} \varphi(a) \lneqq \inf\{\operatorname{Re} \varphi(x);\ x \in C\}.$$

We denote the right-hand side by α. Noting that φ is continuous in the weak topology, we see that $U = \{x \in X;\ \operatorname{Re} \varphi(x) < \alpha\}$ is a neighborhood of a in the weak topology satisfying $U \cap C = \emptyset$. Thus, C is closed in the weak topology. $\qquad\square$

Let X be a linear space, and let $x_1, x_2, \ldots, x_n \in X$. If y is expressed as $y = \sum_{i=1}^{n} t_i x_i$, $t_1, t_2, \ldots, t_n \geq 0$, $\sum_{i=1}^{n} t_i = 1$, we say that y is a *convex combination* of $x_1, x_2, \ldots, x_n$. For a subset A of X, we denote by conv A the set of convex combinations of elements in A and call it the *convex span* of A.

Problem 5.2. Show that the closure $\overline{C}$ of a convex subset C of a topological linear space X is again a convex set.

In what follows, we denote the closure of C in a specific topology σ by $\overline{C}^\sigma$.

Corollary 5.3. *Assume that a sequence $\{x_n\}_{n=1}^{\infty}$ in a normed space X weakly converges to $x \in X$. Then, the following holds: $\forall \varepsilon > 0$, $\exists y \in \operatorname{conv}\{x_n\}_{n=1}^{\infty}$, $\|x - y\| < \varepsilon$.*

Proof. Let $C = \overline{\operatorname{conv}\{x_n\}_{n=1}^{\infty}}^{\|\cdot\|}$. Then, C is a convex set that is closed in the norm topology, and Corollary 5.2 implies that it is closed in the weak topology. Thus, we get $x \in C$. $\qquad\square$

Corollary 5.4. *For a normed space X, the closed unit ball B_X is dense in $B_{X^{**}}$ in the weak* topology $\sigma(X^{**}, X^*)$ of X^{**}.*

Proof.   Note that $\overline{B_X}^{\sigma(X^{**},X^*)} \subset B_{X^{**}}$. Indeed, if $x^{**} \in X^{**}$ satisfies $\|x^{**}\| > 1$, there exists $\varphi \in B_{X^*}$ satisfying $|x^{**}(\varphi)| > 1$, and we can separate x^{**} from B_X by a neighborhood in the weak* topology of X^{**}.

We assume that $y^{**} \in B_{X^{**}} \backslash \overline{B_X}^{\sigma(X^{**},X^*)}$ exists and deduce a contradiction in the following. By the Hahn–Banach separation theorem and Corollary 5.1, there would exist $\varphi \in X^*$ satisfying

$$\sup\{\operatorname{Re}\varphi(x);\ x \in B_X\} \lneqq \operatorname{Re} y^{**}(\varphi).$$

While the left-hand side is $\|\varphi\|$, the right-hand side is bounded by $\|\varphi\|$ above, and we get a contradiction. Thus, there is no such y^{**}.   □

Let X be a normed space. Recall that for a subset Y of X, we define $Y^\perp$ by

$$Y^\perp = \{\varphi \in X^*;\ \forall y \in Y,\ \varphi(y) = 0\}.$$

Here, for a subset Z of X^*, we define

$$Z_\perp = \{x \in X;\ \forall \varphi \in Z,\ \varphi(x) = 0\}.$$

Problem 5.3. For a normed space X, show the following:

(1) If Y is a closed subspace in X, the equality $(Y^\perp)_\perp = Y$ holds.
(2) If Z is weak* closed subspace of X^*, the equality $(Z_\perp)^\perp = Z$ holds.

5.3.2   *Reflexive Banach spaces*

We complete the characterization of reflexive Banach spaces, which was pending in Chapter 4.

Theorem 5.5. *For a Banach space X, the following conditions are equivalent:*

(1) *X is reflexive.*
(2) *B_X is compact in the weak topology.*

Proof.   We have already shown (1) $\Longrightarrow$ (2) in Corollary 3.5, and we show (2) $\Longrightarrow$ (1). Note that the weak topology $\sigma(X, X^*)$ of X is the relative topology of the weak* topology $\sigma(X^{**}, X^*)$ of X^{**}.

Thus, B_X is a compact subset of X^{**} in the weak* topology and, in particular, weak* closed. On the other hand, since Corollary 5.4 states that B_X is dense in $B_{X^{**}}$ in the weak* topology, we get $B_X = B_{X^{**}}$ and $X = X^{**}$. $\qquad\square$

Corollary 5.5. *Assume that Y is a closed subspace of a Banach space X.*

(1) *If X is reflexive, so is Y.*
(2) *If X is reflexive, so is X/Y.*
(3) *If X^* is reflexive, so is X.*

Proof. (1) Corollary 5.2 shows that Y is a closed subset of X in the weak topology. Thus, $B_Y = B_X \cap Y$ shows that B_Y is compact in the weak topology of X. On the other hand, since the Hahn–Banach extension theorem shows that the weak topology of Y is the relative topology of that of X, we conclude that B_Y is compact in the weak topology of Y, and Y is reflexive.

(3) Since X^* is reflexive, so is X^{**}. As X is a closed subspace of X^{**}, (1) implies that X is reflexive.

(2) Since X is reflexive, so is X^*, and (1) implies that $Y^\perp$ is reflexive. Since Theorem 3.3 shows that $(X/Y)^*$ is isometrically isomorphic to $Y^\perp$, (3) implies that X/Y is reflexive too. $\qquad\square$

Corollary 5.6. *For a reflexive Banach space X, its closed unit ball B_X is weakly sequentially compact (that is, every sequence in B_X has a weakly convergent subsequence).*

Proof. Let $\{x_n\}_{n=1}^\infty$ be a sequence in B_X, and let Y be the closed subspace generated by $\{x_n\}_{n=1}^\infty$. Then, Y is separable and reflexive. Thus, B_Y is compact in the weak topology. Since $Y = Y^{**}$, Lemma 3.3 shows that Y^* is separable, and Problem 3.7 shows that the weak topology on B_Y is metrizable. Since B_Y is compact and metrizable in the weak topology, it is sequentially compact. $\qquad\square$

In fact, it follows from the following general theorem that the above property is also one of the equivalent conditions for reflexivity. For the proof, the reader is referred to, for example, Dunford and Schwartz (1988, Theorem V.6.1).

Theorem 5.6 (Eberlein–Smulian theorem). *For a subset A of a Banach space X, the following conditions are equivalent:*

(1) *A is relatively weakly compact.*
(2) *A is relatively weakly sequentially compact.*

As an application of Corollary 5.6, we can generalize the projection theorem for a Hilbert space to a class of Banach spaces, including the L^p-spaces, with $1 < p < \infty$.

Definition 5.4. We say that a Banach space X is *strictly convex* if the following holds: $x, y \in X$, $\|x\| = \|y\| = \|x+y\|/2 = 1 \implies x = y$.

Problem 5.4. Show that the L^p-spaces are strictly convex for all $1 < p < \infty$.

Proposition 5.1. *Let X be a reflexive Banach space, and let C be a closed convex subset of X. Then, for every $x \in X$, there exists $x_0 \in C$ satisfying $\|x - x_0\| = d(x, C)$. If moreover X is strictly convex, then x_0 is unique.*

Proof. Take a sequence $\{y_n\}_{n=1}^{\infty}$ in C such that $\{\|x - y_n\|\}_{n=1}^{\infty}$ converges to $d(x, C)$. Then, since $\{y_n\}_{n=1}^{\infty}$ is a bounded sequence and X is reflexive, there exists a weakly convergent subsequence $\{y_{n_k}\}_{k=1}^{\infty}$. Let $x_0 = \text{w-}\lim_{k \to \infty} y_{n_k}$. Then, Exercise 3.3 implies

$$\|x - x_0\| \leq \liminf_{k \to \infty} \|x - y_{n_k}\| = d(x, C).$$

Corollary 5.2 shows that C is a weakly closed convex set, and we get $x_0 \in C$, and so $\|x - x_0\| = d(x_0, C)$ holds.

Assume that $x_0' \in C$ also satisfies the condition. Since C is a convex set, we see that $2^{-1}(x_0 + x_0')$ also satisfies the condition. This means

$$\|x - x_0\| = \|x - x_0'\| = \left\|\frac{1}{2}(x - x_0 + x - x_0')\right\|.$$

Thus, if X is strictly convex, we get $x_0 = x_0'$. $\qquad\square$

5.3.3   *Krein–Milman theorem*

Finally, we finish this chapter by showing the Krein–Milman theorem, which has applications in various fields.

Definition 5.5. Let C be a convex subset of a locally convex space X.

(1) We say that $x \in C$ is an *extreme point* of C if the following holds: $y, z \in C$, $0 < t < 1$, $x = ty + (1-t)z \in C \implies y = z = x$. We denote by $\mathrm{ex}\, C$ the set of extreme points of C.
(2) We say that a non-empty closed convex subset S of C is a *face* of C if the following holds: $y, z \in C$, $0 < t < 1$, $ty + (1-t)z \in S \implies y, z \in S$.

    An element $x \in C$ is an extreme point of C if and only if the one-point set $\{x\}$ is a face of C. When S is a face of C, an extreme point of S is an extreme point of C. The intersection of arbitrarily many faces is a face as long as it is not empty.

Lemma 5.4. *Let C be a non-empty compact convex subset of a locally convex space. Then, $\mathrm{ex}\, C \neq \emptyset$.*

Proof.   Let $\mathcal{F}$ be the set of all faces of C. Then, $\mathcal{F}$ is not empty as $C \in \mathcal{F}$. By setting $S_2 \leq S_1$ when $S_1, S_2 \in \mathcal{F}$ satisfy $S_1 \subset S_2$, we can make $\mathcal{F}$ an ordered set. We show that $\mathcal{F}$ is an inductively ordered set. Let $\mathcal{F}'$ be an arbitrary totally ordered subset of $\mathcal{F}$. For any finitely many faces $S_1, S_2, \ldots, S_n \in \mathcal{F}'$, there exists j satisfying $\bigcap_{i=1}^{n} S_i = S_j$, and $\{S\}_{S \in \mathcal{F}'}$ is a family of closed subsets of C with the finite intersection property. As C is compact, the intersection $S' = \bigcap_{S \in \mathcal{F}'} S$ is not empty, and it is a face. Thus, S' is an upper bound of $\mathcal{F}'$.

    By Zorn's lemma, there exists a maximal element S_0 in $\mathcal{F}$. We assume $a, b \in S_0$, $a \neq b$, and deduce a contradiction. By the Hahn–Banach separation theorem, there exists a continuous linear functional φ satisfying $\mathrm{Re}\, \varphi(a) < \mathrm{Re}\, \varphi(b)$. Let α be the minimum of $\mathrm{Re}\, \varphi$ on S_0, and set $S_0' = \{x \in S_0;\ \mathrm{Re}\, \varphi(x) = \alpha\}$. Since S_0 is compact and φ is continuous, the set S_0' is not empty, and it is a face. However, we have $b \notin S_0'$, which contradicts the maximality of S_0. Thus, S_0 is a one-point set, and it consists of an extreme point of C. $\qquad\square$

Theorem 5.7 (Krein–Milman theorem). *Let C be a compact convex subset of a locally convex space X. Then,* $\overline{\text{conv}}(\text{ex}\,C) = C$.

Proof. As there is nothing to show in the case of $C = \emptyset$, we assume $C \neq \emptyset$. Let $C_0 = \overline{\text{conv}}(\text{ex}\,C)$. Then, C_0 is a non-empty compact convex subset of C. We assume $a \in C \backslash C_0$ exists and deduce a contradiction. Thanks to the Hahn–Banach separation theorem, there exists a continuous linear functional φ of X satisfying

$$\text{Re}\,\varphi(a) \nleqq \inf\{\text{Re}\,\varphi(x); \ x \in C_0\}.$$

Let α be the minimum of $\text{Re}\,\varphi$ on C, and let $S = \{x \in C \ \text{Re}\,\varphi(x) = \alpha\}$. Since C is compact and $\text{Re}\,\varphi$ is continuous, the set S is not empty, and it is a face. Thus, $\text{ex}\,C \supset \text{ex}\,S \neq \emptyset$. However, this contradicts $S \cap C_0 = \emptyset$, and we get $C = C_0$. $\qquad\square$

When X is a Banach space, the Banach–Alaoglu theorem implies that B_{X^*} is a compact convex set in the weak* topology, and B_{X^*} has sufficiently many extreme points.

Corollary 5.7. *The Banach space $L^1[0,1]$ is not isometrically isomorphic to the dual space of any Banach space.*

Proof. It suffices to show that $B_{L^1[0,1]}$ has no extreme point. Let $f \in B_{L^1[0,1]}$. Since f with $\|f\|_1 < 1$ cannot be an extreme point, we assume $\|f\|_1 = 1$. Let $F(t) = \int_0^t |f(s)|ds$. Then, F is continuous on $[0,1]$, and $F(0) = 0$ and $F(1) = 1$. Thus, by the intermediate value theorem, there exists $c \in (0,1)$ satisfying $F(c) = 1/2$. Let $g = 2\chi_{[0,c]}f$ and $h = 2\chi_{[c,1]}f$. Then, $g, h \in B_{L^1[0,1]}$ and $f = 2^{-1}(g+h)$. As $g \neq h$, we conclude that f is not an extreme point. $\qquad\square$

Exercises

Exercise 5.1
Assume that the topology of a locally convex space X is given by a family of semi-norms, $\{p_\lambda\}_{\lambda \in \Lambda}$. Show that a sequence $\{x_n\}_{n=1}^\infty$ in X converges to $x \in X$ if and only if $\{p_\lambda(x - x_n)\}_{n=1}^\infty$ converges to 0 for every $\lambda \in \Lambda$. If you know the definition of a net (see Section A.1), generalize the above statement to the case of a net.

Exercise 5.2
For $t \in \mathbb{R}$, we define a unitary operator on $L^2(\mathbb{R})$ by $U_t f(s) = f(s-t)$ for $f \in L^2(\mathbb{R})$. Show the following:

(1) The map $t \mapsto U_t$ is continuous in the strong operator topology.
(2) $\{U_t\}_{t \in \mathbb{R}}$ converges to 0 in the weak operator topology as $|t| \to \infty$.

Exercise 5.3
For a compact Hausdorff space Ω, we regard the set of regular Borel probability measures $\mathbf{P}(\Omega)$ on Ω as a subset of $B_{C(\Omega)^*}$.

(1) Show that $\mathbf{P}(\Omega)$ is a compact convex set in the weak* topology.
(2) Show $\operatorname{ex} \mathbf{P}(\Omega) = \{\delta_\omega\}_{\omega \in \Omega}$. Here, δ_ω is the Dirac measure $\int_\Omega f d\delta_\omega = f(\omega)$.

Exercise 5.4
In this exercise, we use the notation in Exercise 3.4(2). For $\mu \in \mathbf{P}_T(\Omega)$, show that the following conditions are equivalent:

(1) $\mu \in \operatorname{ex} \mathbf{P}_T(\Omega)$.
(2) If a Borel measurable set $E \subset \Omega$ satisfies $\mu(E \ominus TE) = 0$, we have either $\mu(E) = 0$ or $\mu(E) = 1$. Here, $E \ominus TE$ means the symmetric difference $(E \backslash TE) \cup (TE \backslash E)$.

We call μ satisfying the above equivalent conditions an *ergodic measure* of T.

Chapter 6

The Spectrum of Bounded Operators

Just as eigenvalues play a central role in linear algebra, the spectrum, a generalization of the set of eigenvalues, plays an important role in operator theory. In this chapter, we introduce the spectrum of a bounded operator (or, more generally, an element in a Banach algebra) and derive its basic properties. Continuous functional calculus is an operation as if to substituting a self-adjoint or unitary operator into a continuous function defined on its spectrum, which is the basis of many arguments in the following chapters.

6.1 Banach Algebras and the Spectrum of Their Elements

Definition 6.1. If a ring $\mathcal{A}$ is a linear space over $\mathbb{C}$ and the product is bilinear, we call $\mathcal{A}$ an *algebra* over $\mathbb{C}$. If $\mathcal{A}$ is an algebra over $\mathbb{C}$ and a Banach space such that $\|ST\| \leq \|S\|\|T\|$ holds for all $S, T \in \mathcal{A}$ (submultiplicativity of norm), we call $\mathcal{A}$ a *Banach algebra*.

Example 6.1. For a Banach space X, the set of bounded operators $\mathbf{B}(X)$ is a Banach algebra.

Example 6.2. For a compact Hausdorff space Ω, the set of continuous functions $C(\Omega)$ is a Banach algebra with the pointwise operations and the norm $\|\cdot\|_\infty$.

Although a Banach algebra is not assumed to have a (multiplicative) unit in general, in this chapter we assume that $\mathcal{A}$ is a Banach algebra with a unit $I_\mathcal{A}$ satisfying $\|I_\mathcal{A}\| = 1$. We often simply denote $I_\mathcal{A}$ by I. We denote by $\mathcal{A}^{-1}$ the (multiplicative) group of all invertible elements in $\mathcal{A}$.

Definition 6.2. For $T \in \mathcal{A}$, we define the following sets:

(1) The *spectrum* $\sigma(T)$ of T is defined to be $\sigma(T) = \{\lambda \in \mathbb{C}; \lambda I - T \notin \mathcal{A}^{-1}\}$.
(2) The *resolvent set* $\rho(T)$ of T is defined to be $\mathbb{C} \backslash \sigma(T)$. For $\lambda \in \rho(T)$, we call $(\lambda I - T)^{-1}$ the *resolvent*.

When X is a Banach space and $T \in \mathbf{B}(X)$, we apply the above definition to $\mathcal{A} = \mathbf{B}(X)$ unless otherwise stated.

For $\lambda, \mu \in \rho(T)$, the following *resolvent identity* holds:

$$
\begin{aligned}
(\lambda I &- T)^{-1} - (\mu I - T)^{-1} \\
&= (\lambda I - T)^{-1} \left((\mu I - T) - (\lambda I - T)\right)(\mu I - T)^{-1} \\
&= (\mu - \lambda)(\lambda I - T)^{-1}(\mu I - T)^{-1}.
\end{aligned}
$$

Lemma 6.1. *If $T \in \mathcal{A}$ satisfies $\|T\| < 1$, we have $I - T \in \mathcal{A}^{-1}$, and*

$$
(I - T)^{-1} = \sum_{n=0}^{\infty} T^n
$$

holds (here, we let $T^0 = I$). The right-hand side is called the Neumann series.

Proof. From the submultiplicativity of the norm, we have $\|T^n\| \leq \|T\|^n$ for all $n \in \mathbb{N}$, and the Neumann series converges. We denote its sum by S. From the continuity of the product, we get $TS = ST = S - I$ and $S = (I - T)^{-1}$. $\qquad\square$

Lemma 6.2. *The set $\mathcal{A}^{-1}$ is open in $\mathcal{A}$, and the map $T \mapsto T^{-1}$ is continuous on $\mathcal{A}^{-1}$.*

Proof. Let $T_0 \in \mathcal{A}^{-1}$. If $T \in \mathcal{A}$ satisfies $\|T - T_0\| < 1/\|T_0^{-1}\|$, we have

$$T = T_0 - (T_0 - T) = T_0 \left(I - T_0^{-1}(T_0 - T)\right),$$

and as $\|T_0^{-1}(T_0 - T)\| \leq \|T_0^{-1}\|\|T_0 - T\| < 1$, we get $T \in \mathcal{A}^{-1}$. Thus, $\mathcal{A}^{-1}$ is an open set. Since

$$T^{-1} - T_0^{-1} = \left(\left(I - T_0^{-1}(T_0 - T)\right)^{-1} - I\right) T_0^{-1},$$

we get the following estimate from the Neumann series:

$$\|T^{-1} - T_0^{-1}\| = \left\|\sum_{n=1}^{\infty} \left(T_0^{-1}(T_0 - T)\right)^n T_0^{-1}\right\|$$

$$\leq \sum_{n=1}^{\infty} \|T_0 - T\|^n \|T_0^{-1}\|^{n+1} = \frac{\|T_0 - T\|\|T_0^{-1}\|^2}{1 - \|T_0 - T\|\|T_0^{-1}\|},$$

and we see that $T \mapsto T^{-1}$ is continuous. $\qquad\qquad\square$

Lemma 6.3. *Let $T \in \mathcal{A}$.*

(1) *The set $\rho(T)$ is open, and the map $\rho(T) \to \mathcal{A}^{-1}$, $\lambda \mapsto (\lambda I - T)^{-1}$, is continuous.*

(2) *For $\lambda \in \rho(T)$, the following holds:*

$$\lim_{\mu \to \lambda} \frac{1}{\mu - \lambda} \left((\mu I - T)^{-1} - (\lambda I - T)^{-1}\right) = -(\lambda I - T)^{-2}.$$

(3) *For $\lambda \in \mathbb{C}$, with $|\lambda| > \|T\|$, we have $\lambda \in \rho(T)$, and the following holds:*

$$\|(\lambda I - T)^{-1}\| \leq \frac{1}{|\lambda| - \|T\|}.$$

Proof. (1) Since the map $\mathbb{C} \to \mathcal{A}$, $\lambda \mapsto \lambda I - T$, is continuous, and $\rho(T)$ is the preimage of $\mathcal{A}^{-1}$ by this map, the set $\rho(T)$ is open in $\mathbb{C}$. Also, since the map $T \mapsto T^{-1}$ is continuous on $\mathcal{A}^{-1}$, so is $\lambda \mapsto (\lambda I - T)^{-1}$ on $\rho(T)$.

(2) From the resolvent identity, we get

$$\frac{1}{\mu - \lambda} \left((\mu I - T)^{-1} - (\lambda I - T)^{-1} \right) = -(\mu I - T)^{-1}(\lambda I - T)^{-1},$$

and the statement follows from the continuity of the resolvent.

(3) Since $\lambda I - T = \lambda(I - \lambda^{-1}T)$ and $\|\lambda^{-1}T\| < 1$, we have $\lambda \in \rho(T)$, and

$$(\lambda I - T)^{-1} = \frac{1}{\lambda}(I - \frac{1}{\lambda}T)^{-1} = \sum_{n=0}^{\infty} \frac{1}{\lambda^{n+1}} T^n.$$

From this, we get the estimate

$$\|(\lambda I - T)^{-1}\| \le \sum_{n=0}^{\infty} \frac{\|T\|^n}{|\lambda|^{n+1}} = \frac{1}{|\lambda| - \|T\|}. \qquad \square$$

Theorem 6.1. *The spectrum $\sigma(T)$ of $T \in \mathcal{A}$ is a non-empty compact subset of $\mathbb{C}$.*

Proof. Since we have already seen, from the above lemma, that $\sigma(T)$ is a bounded closed subset of $\mathbb{C}$, it suffices to show that $\sigma(T)$ is non-empty. We assume $\sigma(T) = \emptyset$ and deduce a contradiction in the following. We take $\varphi \in \mathcal{A}^*$ and set $f(z) = \varphi((zI - T)^{-1})$ for $z \in \rho(T) = \mathbb{C}$. Then, (2) of the previous lemma shows that f is holomorphic. Since we have $|f(z)| \le \|\varphi\|(|z| - \|T\|)^{-1} \le \|\varphi\|$ for $|z| \ge 1 + \|T\|$, which shows that f is bounded on $\mathbb{C}$, the Liouville's theorem implies that it is a constant function. Furthermore, we have $f(z) \to 0$ as $|z| \to \infty$, and we get $f(z) = 0$. For fixed $z \in \mathbb{C}$, we have $\varphi((zI - T)^{-1}) = 0$ for all $\varphi \in \mathcal{A}^*$, and we get $(zI - T)^{-1} = 0$, which is a contradiction. Thus, $\sigma(T) \ne \emptyset$. $\qquad \square$

Lemma 6.4 (Spectral mapping theorem). *For a polynomial $f(z) = \sum_{n=0}^{N} c_n z^n$ and $T \in \mathcal{A}$, we let $f(T) = \sum_{n=0}^{N} c_n T^n$. Then, $\sigma(f(T)) = f(\sigma(T))$ holds.*

Proof. Note that, in general, when $S_1, S_2, \ldots, S_N \in \mathcal{A}$ commute with each other, the product $S_1 S_2 \ldots S_N$ is invertible if and only if each S_i is invertible for $i = 1, 2, \ldots, N$.

We may assume $c_N \neq 0$ for showing the lemma. Fixing $\lambda \in \mathbb{C}$ and decomposing $g(z) = f(z) - \lambda$ as

$$c_N(z - \lambda_1)(z - \lambda_2) \cdots (z - \lambda_N),$$

we get

$$f(T) - \lambda I = c_N(T - \lambda_1 I)(T - \lambda_2 I) \cdots (T - \lambda_N I).$$

Then,

$$\lambda \in \rho(f(T)) \iff f(T) - \lambda I \in \mathcal{A}^{-1} \iff \forall i,\ T - \lambda_i I \in \mathcal{A}^{-1}$$
$$\iff \forall i,\ \lambda_i \in \rho(T),$$

and

$$\lambda \in \sigma(f(T)) \iff \exists i,\ \lambda_i \in \sigma(T) \iff 0 \in g(\sigma(T)) \iff \lambda \in f(\sigma(T)).$$

Therefore, we get $\sigma(f(T)) = f(\sigma(T))$. $\square$

Problem 6.1. For $T \in \mathcal{A}^{-1}$ and the Laurent polynomial $f(z) = \sum_{n=-M}^{N} c_n z^n$, we let $f(T) = \sum_{n=-M}^{N} c_n T^n$. Show $\sigma(f(T)) = f(\sigma(T))$.

Definition 6.3. For $T \in \mathcal{A}$, we define the *spectral radius* of T by $r(T) = \max_{\lambda \in \sigma(T)} |\lambda|$. As $|\lambda| > \|T\|$ implies $\lambda \in \rho(T)$, we have $r(T) \leq \|T\|$.

Theorem 6.2. *For $T \in \mathcal{A}$, the following formula holds:*

$$r(T) = \lim_{n \to \infty} \|T^n\|^{\frac{1}{n}} = \inf_{n \in \mathbb{N}} \|T^n\|^{\frac{1}{n}}.$$

Proof. From the spectral mapping theorem, we have $r(T)^n = r(T^n) \leq \|T^n\|$, and $r(T) \leq \inf_{n \in \mathbb{N}} \|T^n\|^{1/n}$ holds.

Let $z \in \mathbb{C}$, with $|z| > \|T\|$. Since

$$(zI - T)^{-1} = \sum_{n=0}^{\infty} \frac{1}{z^{n+1}} T^n,$$

for every $\varphi \in \mathcal{A}^*$, we have

$$\varphi((zI - T)^{-1}) = \sum_{n=0}^{\infty} \frac{1}{z^{n+1}} \varphi(T^n).$$

Since the left-hand side is holomorphic on $|z| > r(T)$, the uniqueness of the Laurent expansion shows that the equality holds on $|z| > r(T)$, and in particular the series on the right-hand side converges. Thus, for $|z| > r(T)$, we have $\lim_{n\to\infty} \varphi(z^{-n}T^n) = 0$. Since this is the case for every φ, the sequence $\{z^{-n}T^n\}_{n=1}^{\infty}$ weakly converges to 0, and the principle of uniform boundedness implies that there exists a constant $M_z > 0$ such that $\|z^{-n}T^n\| \le M_z$ holds for all $n \in \mathbb{N}$. From this, we can see that

$$\limsup_{n\to\infty} \|T^n\|^{\frac{1}{n}} \le |z|.$$

Since $|z| > r(T)$ is arbitrary, we get

$$r(T) \le \inf_{n\in\mathbb{N}} \|T^n\|^{\frac{1}{n}} \le \liminf_{n\to\infty} \|T^n\|^{\frac{1}{n}} \le \limsup_{n\to\infty} \|T^n\|^{\frac{1}{n}} \le r(T). \qquad \square$$

Problem 6.2. We define $V \in \mathbf{B}(C[0,1])$ by $Vf(t) = \int_0^t f(s)ds$.

(1) For $n \in \mathbb{N}$, show

$$V^n f(t) = \frac{1}{(n-1)!} \int_0^t (t-s)^{n-1} f(s)ds.$$

(2) Calculate the spectral radius $r(V)$ of V.
(3) Determine the spectrum $\sigma(V)$ of V.

6.2 Classification of Points in the Spectrum

Throughout this section, we assume that X is a Banach space, $T \in \mathbf{B}(X)$, and $\lambda \in \mathbb{C}$. If $\lambda \in \rho(T)$, or equivalently $\lambda I - T \in \mathbf{B}(X)^{-1}$, we have $\ker(\lambda I - T) = \{0\}$ and $\mathcal{R}(\lambda I - T) = X$. On the other hand, if $\ker(\lambda I - T) = \{0\}$ and $\mathcal{R}(\lambda I - T) = X$, the Banach inverse mapping theorem implies $\lambda I - T \in \mathbf{B}(X)^{-1}$ and $\lambda \in \rho(T)$. Thus,

$$\lambda \in \sigma(T) \iff \ker(\lambda I - T) \ne \{0\} \text{ or } \mathcal{R}(\lambda I - T) \ne X.$$

From this fact, points in $\sigma(T)$ are classified as follows.

Definition 6.4. The spectrum $\sigma(T)$ is a disjoint union of the following three subsets:

(1) The *point spectrum* $\sigma_{\mathrm{p}}(T)$ of T is defined to be

$$\sigma_{\mathrm{p}}(T) = \{\lambda \in \mathbb{C};\ \ker(\lambda I - T) \neq \{0\}\}.$$

An element $\lambda \in \sigma_{\mathrm{p}}(T)$ is an eigenvalue of T, and $x \in \ker(\lambda I - T)\backslash\{0\}$ is an eigenvector.

(2) The *continuous spectrum* $\sigma_{\mathrm{c}}(T)$ of T is defined to be

$$\sigma_{\mathrm{c}}(T) = \{\lambda \in \mathbb{C};\ \ker(\lambda I - T) = \{0\},$$
$$\mathcal{R}(\lambda I - T) \neq X,\ \overline{\mathcal{R}(\lambda I - T)} = X\}.$$

(3) The *residual spectrum* $\sigma_{\mathrm{r}}(T)$ of T is defined to be

$$\sigma_{\mathrm{r}}(T) = \{\lambda \in \mathbb{C};\ \ker(\lambda I - T) = \{0\},\ \overline{\mathcal{R}(\lambda I - T)} \neq X\}.$$

If X is finite-dimensional, we have $\sigma(T) = \sigma_{\mathrm{p}}(T)$. However, if X is infinite-dimensional, the set $\sigma_{\mathrm{p}}(T)$ may be empty, and we cannot use eigenvalues or eigenvectors for the analysis of T. Nevertheless, eigenvectors in an approximate sense are useful when we work on some operators, such as normal operators on Hilbert spaces.

Definition 6.5. We say that T is *bounded below* if there exists $c > 0$ such that $\|Tx\| \geq c\|x\|$ holds for all $x \in X$. As $T \in \mathbf{B}(X)^{-1}$ implies $\|x\| = \|T^{-1}Tx\| \leq \|T^{-1}\|\|Tx\|$, being bounded below is a necessary condition for T to be invertible.

Definition 6.6. We define the *approximate point spectrum* of T by

$$\sigma_{\mathrm{ap}}(T) = \{\lambda \in \mathbb{C};\ \lambda I - T \text{ is not bounded below}\}.$$

From the definition, a complex number λ belongs to $\sigma_{\mathrm{ap}}(T)$ if and only if there exists a sequence of unit vectors $\{x_n\}_{n=1}^{\infty}$ in X such that $\{\lambda x_n - Tx_n\}_{n=1}^{\infty}$ converges to 0. Such a sequence $\{x_n\}_{n=1}^{\infty}$ is called an *approximate eigenvector*. In particular, we have $\sigma_{\mathrm{p}}(T) \subset \sigma_{\mathrm{ap}}(T)$.

The following lemma will be frequently used in the following section.

Lemma 6.5. *If T is bounded below, its range $\mathcal{R}(T)$ is closed.*

Proof. From the assumption, there exists $c > 0$ such that $\|Tx\| \geq c\|x\|$ holds for all $x \in X$. For $y \in \overline{\mathcal{R}(T)}$, there exists a sequence

$\{x_n\}_{n=1}^\infty$ in X such that $\{Tx_n\}_{n=1}^\infty$ converges to y. Then, $\|Tx_m - Tx_n\| \geq c\|x_m - x_n\|$, and $\{x_n\}_{n=1}^\infty$ is a Cauchy sequence in X. As X is a Banach space, it converges. Letting $\lim_{n\to\infty} x_n = x$, we get $Tx = y$, and $y \in \mathcal{R}(T)$. $\qquad\square$

Corollary 6.1. $\sigma_{\mathrm{p}}(T) \cup \sigma_{\mathrm{c}}(T) \subset \sigma_{\mathrm{ap}}(T)$.

Proposition 6.1. $\partial\sigma(T) \subset \sigma_{\mathrm{ap}}(T)$.

Proof. Let $\lambda \in \partial\sigma(T)$. Then, there exists a sequence $\{\lambda_n\}_{n=1}^\infty$ in $\rho(T)$ converging to λ. On the other hand, $\lambda I - T$ is not invertible as $\partial\sigma(T) \subset \sigma(T)$. Since

$$\lambda I - T = \lambda_n I - T - (\lambda_n - \lambda)I = \{I - (\lambda_n - \lambda)(\lambda_n I - T)^{-1}\}(\lambda_n I - T)$$

is not invertible, we have $\|(\lambda_n - \lambda)(\lambda_n I - T)^{-1}\| \geq 1$, and there exists $x_n \in X$ satisfying $\|x_n\| = 1$ and

$$\|(\lambda_n I - T)^{-1} x_n\| > \frac{1}{2|\lambda_n - \lambda|} \to \infty, \quad (n \to \infty).$$

Let $y_n = (\lambda_n I - T)^{-1} x_n$. Then, we have

$$\left\|(\lambda I - T)\frac{1}{\|y_n\|}y_n\right\| = \left\|(\lambda - \lambda_n)\frac{1}{\|y_n\|}y_n + (\lambda_n I - T)\frac{1}{\|y_n\|}y_n\right\|$$

$$\leq |\lambda_n - \lambda| + \frac{\|x_n\|}{\|y_n\|} \to 0, \quad (n \to \infty),$$

and $\lambda \in \sigma_{\mathrm{ap}}(T)$. $\qquad\square$

Hereafter, we assume that $X = \mathcal{H}$ is a Hilbert space. In this case, since $(ST)^* = T^*S^*$ for $S, T \in \mathbf{B}(\mathcal{H})$, we have $T \in \mathbf{B}(\mathcal{H})^{-1} \iff T^* \in \mathbf{B}(\mathcal{H})^{-1}$, and so $(T^*)^{-1} = (T^{-1})^*$ holds. Since $(\lambda I - T)^* = \overline{\lambda} I - T^*$, we have $\sigma(T^*) = \{\overline{\lambda} \in \mathbb{C};\ \lambda \in \sigma(T)\}$.

Lemma 6.6. *For $T \in \mathbf{B}(\mathcal{H})$, the following hold:*

(1) $\lambda \in \sigma_{\mathrm{r}}(T) \implies \overline{\lambda} \in \sigma_{\mathrm{p}}(T^*)$.
(2) $\lambda \in \sigma_{\mathrm{p}}(T) \implies \overline{\lambda} \in \sigma_{\mathrm{p}}(T^*) \cup \sigma_{\mathrm{r}}(T^*)$.
(3) $\lambda \in \sigma_{\mathrm{c}}(T) \iff \overline{\lambda} \in \sigma_{\mathrm{c}}(T^*)$.

Proof. (1) follows from

$$\lambda \in \sigma_r(T) \Longrightarrow \{0\} \neq \mathcal{R}(\lambda I - T)^{\perp} = \ker((\lambda I - T)^*).$$

(2) follows from

$$\lambda \in \sigma_p(T) \iff \{0\} \neq \ker(\lambda I - T) = \mathcal{R}((\lambda I - T)^*)^{\perp}.$$

From the above results, we get

$$\lambda \in \sigma_p(T) \cup \sigma_r(T) \Longrightarrow \bar{\lambda} \in \sigma_p(T^*) \cup \sigma_r(T^*),$$

and $\bar{\lambda} \in \sigma_c(T^*) \Longrightarrow \lambda \in \sigma_c(T)$ holds. Switching the roles of T and T^*, we get the statement of (3). □

Example 6.3. We determine the spectrum of the unilateral shift V on ℓ^2. Since V is an isometry, so is V^n for every $n \in \mathbb{N}$, and we get $\|V^n\| = 1$. Thus, $r(V) = \lim_{n \to \infty} \|V^n\|^{1/n} = 1$. If $x \in \ell^2 \backslash \{0\}$ and $\lambda \in \mathbb{C}$ satisfy $V^*x = \lambda x$, we have $x_n = \lambda^{n-1}x_1$, and $x \in \ell^2 \backslash \{0\}$ implies $|\lambda| < 1$. On the other hand, if we set $x_n = \lambda^n$ for $|\lambda| < 1$, we have $x = (x_n)_{n=1}^{\infty} \in \ell^2 \backslash \{0\}$, and $V^*x = \lambda x$ holds. Thus, $\sigma_p(V^*) = \{\lambda \in \mathbb{C}; \ |\lambda| < 1\} = \mathbb{D}$. Since $\sigma(V^*)$ is compact and $r(V^*) = r(V) = 1$, we get $\sigma(V^*) = \bar{\mathbb{D}}$. Since $\sigma(V) = \{\bar{\lambda}; \ \lambda \in \sigma(V^*)\}$, we conclude that $\sigma(V) = \bar{\mathbb{D}}$.

Problem 6.3. Determine $\sigma_i(V)$ and $\sigma_i(V^*)$ for $i = p, c, r, ap$.

We derive properties of the spectrum of a normal operator as follows.

Lemma 6.7. *Let $T \in B(\mathcal{H})$ be a normal operator.*

(1) *We have $\|Tx\| = \|T^*x\|$ for all $x \in \mathcal{H}$. In particular, the equality $\ker(\lambda I - T) = \ker(\bar{\lambda}I - T^*)$ holds for all $\lambda \in \mathbb{C}$.*
(2) *If $\lambda, \mu \in \sigma_p(T)$, $\lambda \neq \mu$, we have $\ker(\lambda I - T) \perp \ker(\mu I - T)$.*

Proof. (1) The first statement follows from

$$\|Tx\|^2 = \langle T^*Tx, x \rangle = \langle TT^*x, x \rangle = \|T^*x\|^2.$$

Since $\lambda I - T$ is normal too, the second statement follows.

(2) For $x \in \ker(\lambda I - T)$ and $y \in \ker(\mu I - T)$, we get

$$\lambda \langle x, y \rangle = \langle Tx, y \rangle = \langle x, T^*y \rangle = \langle x, \overline{\mu} y \rangle = \mu \langle x, y \rangle,$$

which shows $x \perp y$. □

Theorem 6.3. *For a normal operator, we have* $\sigma(T) = \sigma_{\mathrm{ap}}(T)$ *and* $\sigma_{\mathrm{r}}(T) = \emptyset$.

Proof. Since $\sigma_{\mathrm{p}}(T) \cup \sigma_{\mathrm{c}}(T) \subset \sigma_{\mathrm{ap}}(T)$ holds in general, it suffices to show $\sigma_{\mathrm{r}}(T) = \emptyset$. Assume $\lambda \in \sigma_{\mathrm{r}}(T)$. Then, we would have $\ker(\lambda I - T) = \{0\}$ and

$$\{0\} \neq \mathcal{R}(\lambda I - T)^{\perp} = \ker(\overline{\lambda} I - T^*) = \ker(\lambda I - T),$$

which is a contradiction. Thus, $\sigma_{\mathrm{r}}(T) = \emptyset$. □

Theorem 6.4. *For a normal operator* $T \in \mathbf{B}(\mathcal{H})$, *we have* $r(T) = \|T\|$.

Proof. Recall Proposition 2.2(2). Let $A = T^*T$. Then, it is self-adjoint, and

$$\|T^{2^n}\|^2 = \|(T^{2^n})^* T^{2^n}\| = \|A^{2^n}\| = \|(A^{2^{n-1}})^* A^{2^{n-1}}\| = \|A^{2^{n-1}}\|^2$$

$$= \cdots = \|A\|^{2^n} = \|T\|^{2^{n+1}}$$

holds. Thus, $r(T) = \lim_{n \to \infty} \|T^{2^n}\|^{\frac{1}{2^n}} = \|T\|$. □

Example 6.4. We have seen that the multiplication operator $M_f \in \mathbf{B}(L^2[0,1])$ of $f \in C[0,1]$ is a normal operator and $\|M_f\| = \|f\|_{\infty}$ holds. We show $\sigma(M_f) = f([0,1])$ here. For $\lambda \notin f([0,1])$, the function $g(t) = (\lambda - f(t))^{-1}$ is continuous on $[0,1]$, and $(\lambda I - M_f)M_g = M_g(\lambda I - M_f) = I$ holds, which shows $\lambda \in \rho(M_f)$. For $\lambda \in f([0,1])$, we can take $t_0 \in [0,1]$ satisfying $f(t_0) = \lambda$. For this t_0, we choose h_n as in Example 1.6. Then, $\|h_n\|_2 = 1$, and

$$\|\lambda h_n - M_f h_n\|_2^2 = \frac{1}{|I_n|} \int_{I_n} |f(t_0) - f(t)|^2 dt$$

$$\leq \max_{t \in I_n} |f(t_0) - f(t)|^2 \to 0, \quad (n \to \infty),$$

which shows $\lambda \in \sigma_{\mathrm{ap}}(M_f)$.

6.3 Continuous Functional Calculus

Throughout this and the following sections, we assume that $\mathcal{H}$ is a Hilbert space.

Lemma 6.8. *For $A \in \mathbf{B}(\mathcal{H})_{\mathrm{sa}}$, we have $\sigma(A) \subset \mathbb{R}$.*

Proof. Note that we have $\sigma(A) = \sigma_{\mathrm{ap}}(A)$ because A is normal. Let $\lambda = \xi + i\eta$, $\xi, \eta \in \mathbb{R}$, $\eta \neq 0$, and let $B = \xi I - A \in \mathbf{B}(\mathcal{H})_{\mathrm{sa}}$. For $x \in \mathcal{H}$,

$$\|(\lambda I - A)x\|^2 = \|Bx + \eta i x\|^2 = \|Bx\|^2 - \eta i \langle Bx, x \rangle$$
$$+ \eta i \langle x, Bx \rangle + \eta^2 \|x\|^2$$
$$= \|Bx\|^2 + \eta^2 \|x\|^2 \geq \eta^2 \|x\|^2,$$

and $\lambda I - A$ is bounded below. Thus, $\lambda \notin \sigma_{\mathrm{ap}}(A)$. $\qquad\square$

Let Ω be a compact subset of $\mathbb{R}$, and let $C_p(\Omega)$ be the set of polynomial functions restricted to Ω. Then, $C_p(\Omega)$ is a dense subalgebra of $C(\Omega)$. To see this, first we choose a finite closed interval $[a, b]$, with $\Omega \subset [a, b]$, and extend the elements in $C(\Omega)$ to continuous functions on $[a, b]$ by applying the Tietze extension theorem to them. By the Weierstrass polynomial approximation theorem, every continuous function on $[a, b]$ can be uniformly approximated by polynomials, and we get the claim (or we can directly apply the Stone–Weierstrass theorem, Theorem A.4). In the following argument, in order to emphasize that we consider the restriction of polynomials to Ω, we denote the norm $\| \cdot \|_\infty$ of $C(\Omega)$ by $\| \cdot \|_{C(\Omega)}$.

Lemma 6.9. *For $A \in \mathbf{B}(\mathcal{H})_{\mathrm{sa}}$ and a polynomial $f(t)$ (respectively, with real coefficients), the operator $f(A)$ is normal (respectively, selfadjoint), and $\|f(A)\| = \|f\|_{C(\sigma(A))}$ holds.*

Proof. We show only the second part, as the first part is straightforward. Since $f(A)$ is normal, we have $\|f(A)\| = r(f(A))$. On the other hand, the spectral mapping theorem shows $\sigma(f(A)) = f(\sigma(A))$, and we get $r(f(A)) = \max_{t \in \sigma(A)} |f(t)|$. $\qquad\square$

Theorem 6.5 (Continuous functional calculus). *For $A \in \mathbf{B}(\mathcal{H})_{\mathrm{sa}}$, there exists a unique algebra homomorphism over $\mathbb{C}$, Φ : $C(\sigma(A)) \to \mathbf{B}(\mathcal{H})$ satisfying the following properties:*

(1) $\Phi(1) = I$.
(2) $\Phi(t) = A$, *where* $t : \sigma(A) \hookrightarrow \mathbb{R}$ *is the embedding map.*
(3) Φ *is an isometry.*

Proof. If Φ exists, (1) and (2) show that it is uniquely determined on $C_p(\sigma(A))$. Since $C_p(\sigma(A))$ is dense in $C(\sigma(A))$, (3) shows that Φ is unique.

To show the existence of Φ, first we would like to define $\Phi_0(f) = f(A)$ for a polynomial $f(t)$. Assume that $f_1(t)$ and $f_2(t)$ are polynomials with $f_1(t) = f_2(t)$ on $\sigma(A)$. Letting $g(t) = f_1(t) - f_2(t)$, we get

$$\|f_1(A) - f_2(A)\| = \|g(A)\| = \|g\|_{C(\sigma(A))} = 0,$$

and $f_1(A) = f_2(A)$. Thus, we can define an algebra homomorphism over $\mathbb{C}$, $\Phi_0 : C_p(\sigma(A)) \to \mathbf{B}(\mathcal{H})$, and the previous lemma shows that it is an isometry. Theorem 1.3 implies that Φ_0 extends to an isometry: $\Phi : C(\sigma(A)) \to \mathbf{B}(\mathcal{H})$.

Since Φ is linear by construction, to complete the proof, it suffices to show that Φ is multiplicative. For arbitrary $f, g \in C(\sigma(A))$, we choose polynomial sequences $\{f_n\}_{n=1}^{\infty}$ and $\{g_n\}_{n=1}^{\infty}$ such that they uniformly converge to f and g, respectively, on $\sigma(A)$. Let $h_n(t) = f_n(t)g_n(t)$. Then, $\{h_n\}_{n=1}^{\infty}$ uniformly converges to fg on $\sigma(A)$, and we get

$$\Phi(fg) = \lim_{n \to \infty} \Phi_0(f_n g_n) = \lim_{n \to \infty} \Phi_0(f_n)\Phi_0(g_n) = \Phi(f)\Phi(g). \qquad \square$$

For $f \in C(\sigma(A))$, we denote $\Phi(f)$ by $f(A)$.

Theorem 6.6. *For $A \in \mathbf{B}(\mathcal{H})_{\mathrm{sa}}$ and $f \in C(\sigma(A))$, the following hold:*

(1) $f(A)$ *is normal.*
(2) $f(A)^* = \overline{f}(A)$ *holds. Here,* $\overline{f}(t) = \overline{f(t)}$. *In particular, if f is real-valued, we have* $f(A) \in \mathbf{B}(\mathcal{H})_{\mathrm{sa}}$.

(3) *If* $\{x_n\}_{n=1}^{\infty}$ *is an approximate eigenvector of* A *associated with* $\lambda \in \sigma(A)$, *the sequence* $\{x_n\}_{n=1}^{\infty}$ *is an approximate eigenvector of* $f(A)$ *associated with* $f(\lambda)$. *In particular, if* $Ax = \lambda x$ *holds, we have* $f(A)x = f(\lambda)x$.

(4) $\sigma(f(A)) = f(\sigma(A))$ *holds (spectral mapping theorem).*

(5) *If* $T \in \mathbf{B}(\mathcal{H})$ *satisfies* $AT = TA$, *we have* $f(A)T = Tf(A)$.

Proof. We take a polynomial sequence $\{f_n\}_{n=1}^{\infty}$ uniformly converging to f on $\sigma(A)$.

(1) Since each $f_n(A)$ is normal and $\{f_n(A)\}_{n=1}^{\infty}$ converges to $f(A)$ in norm, we see that $f(A)$ is normal.

(2) Since $f_n(A)^* = \overline{f_n}(A)$, and $\{\overline{f_n}\}_{n=1}^{\infty}$ uniformly converges to $\overline{f}$ on $\sigma(A)$, we get

$$f(A)^* = \lim_{n \to \infty} f_n(A)^* = \lim_{n \to \infty} \overline{f_n}(A) = \overline{f}(A).$$

(3) Since we have

$$\|(A^{k+1} - \lambda^{k+1}I)x_n\| = \|A(A^k - \lambda^k I)x_n + \lambda^k(A - \lambda I)x_n\|$$
$$\leq \|A\|\|(A^k - \lambda^k I)x_n\| + |\lambda^k|\|(A - \lambda I)x_n\|$$

for every $k \in \mathbb{N}$, an inductive argument shows that the statement holds for every monomial and hence for every polynomial. Since $\{f_m\}_{m=1}^{\infty}$ uniformly converges to f on $\sigma(A)$, we have the following: $\forall \varepsilon > 0, \exists M \in \mathbb{N}, \forall m \geq M, \|f - f_m\|_{C(\sigma(A))} < \varepsilon$. Also, $\exists N \in \mathbb{N}$, $\forall n \geq N, \|(f_M(A) - f_M(\lambda))x_n\| < \varepsilon$. Thus, $n \geq N$ implies

$$\|(f(A) - f(\lambda))x_n\| \leq \|(f(A) - f_M(A))x_n\| + \|(f_M(A) - f_M(\lambda))x_n\|$$
$$+ \|(f_M(\lambda) - f(\lambda))x_n\|$$
$$\leq \|f(A) - f_M(A)\| + \varepsilon + |f_M(\lambda) - f(\lambda)|$$
$$\leq 3\varepsilon.$$

Thus, $\{(f(A) - f(\lambda))x_n\}_{n=1}^{\infty}$ converges to 0.

(4) Note that $\sigma(A) = \sigma_{\mathrm{ap}}(A)$ and $\sigma(f(A)) = \sigma_{\mathrm{ap}}(f(A))$ hold. From (3), we get $f(\sigma(A)) \subset \sigma(f(A))$. Let $\lambda \in \mathbb{C}\backslash f(\sigma(A))$. Since $g(t) = (\lambda - f(t))^{-1}$ is continuous on $\sigma(A)$ and $(\lambda - f(t))g(t) = 1$,

we get

$$(\lambda I - f(A))g(A) = g(A)(\lambda I - f(A)) = I.$$

Thus, $\lambda \in \rho(f(A))$ holds.

(5) Since $f_n(A)T = Tf_n(A)$, we get $f(A)T = Tf(A)$. □

Example 6.5. Let $\mathcal{H} = L^2[0,1]$, and let $Af(t) = tf(t)$. We have already seen that $\sigma(A) = [0,1]$ in Example 6.4. For $f \in C(\sigma(A)) = C[0,1]$, the operator $f(A)$ is nothing but the multiplication operator M_f of f.

We can apply the continuous functional calculus to unitary operators too.

Lemma 6.10. *If $U \in \mathcal{U}(\mathcal{H})$, we have $\sigma(U) \subset S^1 = \{z \in \mathbb{C}; |z| = 1\}$.*

Proof. Since $\|U\| \leq 1$, we have $\sigma(U) \subset \overline{\mathbb{D}}$ and $\sigma(U^*) \subset \overline{\mathbb{D}}$. Since $U^{-1} = U^*$ and $\sigma(U^{-1}) = \{\lambda^{-1} \in \mathbb{C}; \lambda \in \sigma(U)\}$, thanks to Problem 6.1, we get $\sigma(U) \subset S^1$. □

Using Problem 6.1 and the trigonometric polynomial approximation of the continuous functions on S^1, we can get the following theorem in the same way as in the case of self-adjoint operators.

Theorem 6.7 (Continuous functional calculus). *For $U \in \mathcal{U}(\mathcal{H})$, there exists a unique algebra homomorphism over $\mathbb{C}$, Ψ : $C(\sigma(U)) \to \mathbf{B}(\mathcal{H})$ satisfying the following:*

(1) $\Psi(1) = I$.
(2) $\Psi(z) = U$. Here, $z : \sigma(U) \hookrightarrow S^1$ is the embedding map.
(3) Ψ is an isometry.

For $f \in C(\sigma(U))$, we denote $\Psi(f)$ by $f(U)$. We can show the statements corresponding to Theorem 6.6 in the same way.

6.4 Applications of the Continuous Functional Calculus

6.4.1 *Positive operators*

Definition 6.7. We say that $A \in \mathbf{B}(\mathcal{H})_{\text{sa}}$ is a *positive operator* if $\sigma(A) \subset [0, \infty)$. We denote by $\mathbf{B}(\mathcal{H})_+$ the set of positive operators on $\mathcal{H}$.

Theorem 6.8. *For every* $A \in \mathbf{B}(\mathcal{H})_+$, *there exists unique* $B \in \mathbf{B}(\mathcal{H})_+$ *satisfying* $A = B^2$.

Proof. Since $f(t) = \sqrt{t}$ is continuous on $\sigma(A)$, we can set $B = f(A)$, which satisfies $A = B^2$. Since $f(\sigma(A)) \subset [0, \infty)$, the spectral mapping theorem shows $B \in \mathbf{B}(\mathcal{H})_+$. Assume that $B_1 \in \mathbf{B}(\mathcal{H})_+$ also satisfies $A = B_1^2$. From the spectral mapping theorem, we have $\sigma(A) = \{\lambda^2; \ \lambda \in \sigma(B_1)\}$, and $\sigma(B_1) \subset [0, \sqrt{r(A)}]$. We choose a sequence of polynomials with real coefficients, $\{f_n(t)\}_{n=1}^{\infty}$, uniformly converging to $\sqrt{t}$ on $[0, \|A\|]$, and set $g_n(t) = f_n(t^2)$. Then, $\{g_n(t)\}_{n=1}^{\infty}$ uniformly converges to t on $[0, \sqrt{\|A\|}]$. Thus, $\lim_{n \to \infty} \|g_n(B_1) - B_1\| = 0$. On the other hand, as we have $g_n(B_1) = f_n(A)$, we get $B = B_1$. $\qquad \square$

We denote B as above by $\sqrt{A}$, or $A^{1/2}$, and call it the (positive) *square root operator* of A.

Theorem 6.9. *For* $A \in \mathbf{B}(\mathcal{H})_{\mathrm{sa}}$, *the following conditions are equivalent*:

(1) $A \in \mathbf{B}(\mathcal{H})_+$.
(2) *There exists* $T \in \mathbf{B}(\mathcal{H})$ *satisfying* $A = T^*T$.
(3) $\langle Ax, x \rangle \geq 0$ *holds for every* $x \in \mathcal{H}$.

Proof. We have already shown $(1) \implies (2)$.
 $(2) \implies (3)$. We have $\langle Ax, x \rangle = \langle T^*Tx, x \rangle = \|Tx\|^2 \geq 0$.
 $(3) \implies (1)$. Let $\lambda \in \sigma(A)$. Since $\sigma(A) = \sigma_{\mathrm{ap}}(A)$, we can take an approximate eigenvector $\{x_n\}_{n=1}^{\infty}$ of A associated with λ, and we get

$$|\langle \lambda x_n - Ax_n, x_n \rangle| \leq \|\lambda x_n - Ax_n\| \|x_n\| \to 0.$$

Thus, $\lambda = \lim_{n \to \infty} \langle Ax_n, x_n \rangle \geq 0$. $\qquad \square$

Condition (3) is the most convenient criterion and is often adopted as the definition of positive operators. In this case, we can weaken the assumption $A \in \mathbf{B}(\mathcal{H})_{\mathrm{sa}}$ as $A \in \mathbf{B}(\mathcal{H})$. From (3), we can see that if $A, B \in \mathbf{B}(\mathcal{H})_+$, we have $A + B \in \mathbf{B}(\mathcal{H})_+$.

For $A, B \in \mathbf{B}(\mathcal{H})_{\mathrm{sa}}$, we define $A \leq B$ if $B - A \in \mathbf{B}(\mathcal{H})_+$. Then, $(\mathbf{B}(\mathcal{H})_{\mathrm{sa}}, \leq)$ is an ordered set. Indeed, if $A \leq B$ and $B \leq A$, we have $A - B \in \mathbf{B}(\mathcal{H})_{\mathrm{sa}}$ and $\sigma(A - B) = \{0\}$, and so $\|A - B\| = r(A - B) = 0$. Also, if $A \leq B$ and $B \leq C$, we have $C - A = (C - B) + (B - A) \in \mathbf{B}(\mathcal{H})_+$, and $A \leq C$.

Corollary 6.2. *For $A \in \mathbf{B}(\mathcal{H})_{\mathrm{sa}}$, let $M = \max \sigma(A)$ and $m = \min \sigma(A)$. Then, the following hold:*

$$M = \sup_{\|x\|=1} \langle Ax, x \rangle, \quad m = \inf_{\|x\|=1} \langle Ax, x \rangle.$$

In particular, we have $\|A\| = \sup_{\|x\|=1} |\langle Ax, x \rangle|$.

Proof. By the spectral mapping theorem, we have $MI - A \in \mathbf{B}(\mathcal{H})_+$, and $\langle (MI - A)x, x \rangle \geq 0$ for all $x \in \mathcal{H}$. In particular, if $\|x\| = 1$, we get $\langle Ax, x \rangle \leq M$. On the other hand, since $M \in \sigma(A) = \sigma_{\mathrm{ap}}(A)$, there exists an approximate eigenvector $\{x_n\}_{n=1}^{\infty}$ of A associated with M, and $\langle Ax_n, x_n \rangle \to M$, $(n \to \infty)$, holds. The same argument works for m.

The last statement follows from $\|A\| = r(A) = \max\{|m|, |M|\}$. $\square$

Corollary 6.3. *For $A, B \in \mathbf{B}(\mathcal{H})_+$, the condition $A \leq B$ implies $\|A\| \leq \|B\|$.*

6.4.2 *The polar decomposition of operators*

Let $\mathcal{H}_1, \mathcal{H}_2$ be Hilbert spaces. For $T \in \mathbf{B}(\mathcal{H}_1, \mathcal{H}_2)$, we set $|T| = (T^*T)^{\frac{1}{2}}$ and call it the *absolute value operator* of T. We see that T has polar decomposition as follows.

Definition 6.8. We say that $W \in \mathbf{B}(\mathcal{H}_1, \mathcal{H}_2)$ is a *partial isometry* if $\|Wx\| = \|x\|$ holds for all $x \in \ker W^{\perp}$.

Note that $\ker T = \ker T^*T$ holds for every $T \in \mathbf{B}(\mathcal{H}_1, \mathcal{H}_2)$ in general. Indeed, the inclusion $\ker T \subset \ker T^*T$ is trivial, and for $x \in \ker T^*T$, we have

$$0 = \langle T^*Tx, x \rangle = \langle Tx, Tx \rangle = \|Tx\|^2.$$

Lemma 6.11. *For $W \in \mathbf{B}(\mathcal{H}_1, \mathcal{H}_2)$, the following conditions are equivalent:*

(1) *W is a partial isometry.* (1′) *W^* is a partial isometry.*
(2) *$W^*W \in \mathcal{P}(\mathcal{H}_1)$.* (2′) *$WW^* \in \mathcal{P}(\mathcal{H}_2)$.*
(3) *$W = WW^*W$.* (3′) *$W^* = W^*WW^*$.*

Proof. We show the equivalence of (1), (2), and (3) as follows. The equivalence of (1′), (2′), and (3′) follows similarly and that of (3) and (3′) can be shown by taking the adjoint. We write $P = W^*W$ and $Q = P_{\ker W^\perp}$.

(1) $\Longrightarrow$ (2). We decompose a given $x \in \mathcal{H}_1$ as $x = x_1 + x_2$, $x_1 \in \ker W$, and $x_2 \in \ker W^\perp$. Then,

$$\langle Px, x \rangle = \|Wx\|^2 = \|Wx_2\|^2 = \|x_2\|^2 = \langle Qx, x \rangle,$$

and $P = Q \in \mathcal{P}(\mathcal{H}_1)$ follows.

(2) $\Longrightarrow$ (1). Since P and Q are projections, and

$$\ker P = \ker W^*W = \ker W = \ker Q,$$

we get $P = Q$. For $x \in \ker W^\perp$, we get

$$\|Wx\|^2 = \langle Px, x \rangle = \langle Qx, x \rangle = \|x\|^2.$$

(2) $\Longrightarrow$ (3). Let $T = W - WW^*W = W(I - P)$. From the C*-condition (see Proposition 2.2(2)), we have

$$\|T\|^2 = \|T^*T\| = \|(I - P)P(I - P)\| = 0,$$

and (3) follows

(3) $\Longrightarrow$ (2). Since $0 = W^*(W - WW^*W) = P - P^2$, the operator P is a projection. $\qquad\square$

Note that when W is a partial isometry, the restriction of W to $\ker W^\perp$ is an isometry, and $\mathcal{R}(W) = W(\ker W^\perp)$ is a closed subspace, which coincides with $\ker W^{*\perp}$. We call $W^*W = P_{\ker W^\perp} = P_{\mathcal{R}(W^*)}$ the *initial projection* of W, and we call $WW^* = P_{\mathcal{R}(W)} = P_{\ker W^{*\perp}}$ the *final projection* of W.

Theorem 6.10. *For every $T \in \mathbf{B}(\mathcal{H}_1, \mathcal{H}_2)$, there exists a unique partial isometry $W \in \mathbf{B}(\mathcal{H}_1, \mathcal{H}_2)$ satisfying $\ker T = \ker W$ and $T = W|T|$. Moreover, $W^*T = |T|$ and $|T^*| = W|T|W^*$ hold.*

Proof. Note that for every $x \in \mathcal{H}_1$, we have

$$\|Tx\|^2 = \langle T^*Tx, x \rangle = \langle |T|^2 x, x \rangle = \||T|x\|^2.$$

In particular, $\ker T = \ker |T|$ holds. Let $Q : \mathcal{H}_1 \to \mathcal{H}_1/\ker T = \mathcal{H}_1/\ker |T|$ be the quotient map. Then, the fundamental theorem of homomorphisms shows that there exist linear bijections $\widetilde{T} : \mathcal{H}_1/\ker T \to \mathcal{R}(T)$ and $\widetilde{|T|} : \mathcal{H}_1/\ker T \to \mathcal{R}(|T|)$ satisfying $T = \widetilde{T} \circ Q$ and $|T| = \widetilde{|T|} \circ Q$:

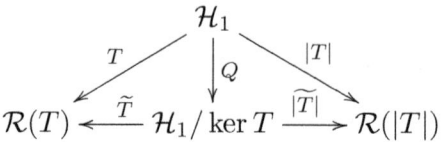

Letting $W_0 = \widetilde{T} \circ \widetilde{|T|}^{-1} : \mathcal{R}(|T|) \to \mathcal{R}(T)$, we get $T = W_0|T|$. Since $\|Tx\| = \||T|x\|$ holds for every $x \in \mathcal{H}_1$, the operator W_0 is an isometry. Note that $\overline{\mathcal{R}(|T|)} = \ker |T|^\perp = \ker T^\perp$. Let $W_1 : \overline{\mathcal{R}(|T|)} \to \overline{\mathcal{R}(T)}$ be the isometric extension of W_0, and let $W = W_1 P_{\ker T^\perp}$. Then, W is a partial isometry satisfying the condition. As W is uniquely determined on $\mathcal{R}(|T|)$ and $\mathcal{R}(|T|)$ is dense in $\ker T^\perp$, the uniqueness of W follows.

The inclusion relation $\mathcal{R}(|T|) \subset \ker T^\perp$ implies $|T| = W^*W|T| = W^*T$. Since $W|T|W^*$ is a positive operator and

$$(W|T|W^*)^2 = W|T|W^*W|T|W^* = W|T|^2W^* = TT^*,$$

we get $W|T|W^* = |T^*|$. $\qquad\square$

We call the above decomposition $T = W|T|$ the *polar decomposition* of T. Since

$$T^* = |T|W^* = W^*W|T|W^* = W^*|T^*|,$$

and the initial projection of W^* is $P_{\overline{\mathcal{R}(T)}} = P_{\ker T^{*\perp}}$, the polar decomposition of T^* is given by $T^* = W^*|T^*|$.

6.4.3 Spectral decomposition (when the spectrum has at most one accumulation point)

First, we summarize basic properties of projections. Note that when $P \in \mathcal{P}(\mathcal{H})$, we have $0 \leq P \leq I$ and, in particular, $\mathcal{P}(\mathcal{H}) \subset \mathbf{B}(\mathcal{H})_+$. Also, since P is the orthogonal projection onto $\mathcal{R}(P)$, we have $\|x\|^2 = \langle Px, x \rangle \iff x \in \mathcal{R}(P)$.

Proposition 6.2. *For $P \in \mathbf{B}(\mathcal{H})_{\mathrm{sa}}$, the following holds:*

$$P \in \mathcal{P}(\mathcal{H}) \iff \sigma(P) \subset \{0, 1\}.$$

Proof. Let $f(t) = t^2 - t$. The spectral mapping theorem shows

$$\sigma(P) \subset \{0, 1\} \iff f(P) = 0 \iff P \in \mathcal{P}(\mathcal{H}). \qquad \square$$

Lemma 6.12. *For $P, Q \in \mathcal{P}(\mathcal{H})$, the following conditions are equivalent:*

(1) $\mathcal{R}(P) \perp \mathcal{R}(Q)$.
(2) $PQ = 0$.
(3) $QP = 0$.
(4) $P + Q \in \mathcal{P}(\mathcal{H})$.

Proof. $(2) \iff (3)$ follows from $(PQ)^* = QP$.
$(2) \iff (1)$ follows from

$$\mathcal{R}(P) \perp \mathcal{R}(Q) \iff \mathcal{R}(Q) \subset \mathcal{R}(P)^{\perp} \iff \mathcal{R}(Q) \subset \ker P$$
$$\iff PQ = 0.$$

$(2) \implies (4)$. Since $P + Q \in \mathbf{B}(\mathcal{H})_{\mathrm{sa}}$ and

$$(P + Q)^2 = P^2 + PQ + QP + Q^2 = P + Q,$$

we get $P + Q \in \mathcal{P}(\mathcal{H})$.
$(4) \implies (2)$. From $(P + Q)^2 = P + Q$, we have $PQ + QP = 0$. From $0 = (PQ + QP)P = P(PQ + QP)$, we have $PQ = QP$, and so $PQ = 0$. $\qquad \square$

When $P, Q \in \mathcal{P}(\mathcal{H})$ satisfy the above equivalent conditions, we say that P and Q are *orthogonal* and write $P \perp Q$.

Lemma 6.13. *For $P, Q \in \mathcal{P}(\mathcal{H})$, the following conditions are equivalent:*

(1) $\mathcal{R}(P) \subset \mathcal{R}(Q)$.
(2) $QP = P$.
(3) $PQ = P$.
(4) $P \leq Q$.

Proof.   (1) $\Longrightarrow$ (2). Since $Px \in \mathcal{R}(Q)$ holds for all $x \in \mathcal{H}$, we get $QPx = Px$.

(2) $\Longleftrightarrow$ (3) follows from $(QP)^* = PQ$.

(3) $\Longrightarrow$ (4). For every $x \in \mathcal{H}$, we have

$$\langle Px, x \rangle = \|Px\|^2 = \|PQx\|^2 \le \|P\|^2 \|Qx\|^2 = \langle Qx, x \rangle,$$

and $P \le Q$ holds.

(4) $\Longrightarrow$ (1). For $x \in \mathcal{R}(P)$, we have

$$\|x\|^2 = \langle Px, x \rangle \le \langle Qx, x \rangle \le \|x\|^2,$$

and $\langle Qx, x \rangle = \|x\|^2$ holds. Thus, $x \in \mathcal{R}(Q)$. $\square$

If $A \in \mathbf{B}(\mathcal{H})_{\mathrm{sa}}$ and $\lambda \in \sigma(A)$ is an isolated point, the indicator function $\chi_{\{\lambda\}}$ of the one-point set $\{\lambda\}$ is continuous on $\sigma(A)$, and $P_\lambda = \chi_{\{\lambda\}}(A)$ is defined. Since $\chi_{\{\lambda\}} = \chi_{\{\lambda\}}^2 = \overline{\chi_{\{\lambda\}}}$, the operator P_λ is a projection. If $\mu \in \sigma(A)$ is an isolated point and $\lambda \ne \mu$, we have $\chi_{\{\lambda\}}\chi_{\{\mu\}} = 0$, and $P_\lambda \perp P_\mu$ holds.

Lemma 6.14. *The projection P_λ is the orthogonal projection onto the eigenspace $\ker(\lambda I - A)$.*

Proof.   Since $t\chi_{\{\lambda\}}(t) = \lambda\chi_{\{\lambda\}}(t)$, we have $AP_\lambda = \lambda P_\lambda$, and $\mathcal{R}(P_\lambda) \subset \ker(\lambda I - A)$ holds. On the other hand, Theorem 6.6(3) shows that $x \in \ker(\lambda I - A)$ implies $P_\lambda x = x$, and the converse inclusion holds. $\square$

Proposition 6.3. *Assume that $A \in \mathbf{B}(\mathcal{H})_{\mathrm{sa}}$ and $\sigma(A)$ is a finite set $\{\lambda_1, \lambda_2, \ldots, \lambda_n\}$. Then, $P_{\lambda_1}, P_{\lambda_2}, \ldots, P_{\lambda_n}$ are mutually orthogonal projections, and the following hold:*

$$I = \sum_{i=1}^{n} P_{\lambda_i}, \quad A = \sum_{i=1}^{n} \lambda_i P_{\lambda_i}.$$

Proof.   Since $1 = \sum_{i=1}^{n} \chi_{\{\lambda_i\}}(t)$ and $t = \sum_{i=1}^{n} \lambda_i \chi_{\{\lambda_i\}}(t)$ hold on $\sigma(A)$, the statement follows. $\square$

The following theorem leads to the spectral decomposition of a compact self-adjoint operator.

Theorem 6.11. *Let $A \in \mathbf{B}(\mathcal{H})_{\text{sa}}$, and assume that $\sigma(A)$ is an infinite set having the only accumulation point 0. Then, we can express $\sigma(A)$ as $\sigma(A) = \{\lambda_n\}_{n=1}^{\infty} \cup \{0\}$; moreover, we may arrange it so that $\{|\lambda_n|\}_{n=1}^{\infty}$ is monotone decreasing and converges to 0, and*

$$A = \sum_{n=1}^{\infty} \lambda_n P_{\lambda_n}$$

converges in norm.

Proof.   Since $\{\lambda \in \sigma(A); |\lambda| \geq 1/n\}$ is a compact set without an accumulation point for $n \in \mathbb{N}$, it is a finite set. Thus, $\sigma(A)$ is a countable set, as in the statement. From

$$\max_{t \in \sigma(A)} \left| t - \sum_{n=1}^{N} \lambda_n \chi_{\{\lambda_n\}}(t) \right| = |\lambda_{N+1}|,$$

we get

$$\left\| A - \sum_{n=1}^{N} \lambda_n P_{\lambda_n} \right\| = |\lambda_{N+1}| \to 0, \quad (N \to \infty). \qquad \square$$

Exercises

Exercise 6.1
Let U be the bilateral shift on $\ell^2(\mathbb{Z})$, and let $A = 2^{-1}(U + U^*)$. Find $\sigma(A)$, and compute $\langle (zI - A)^{-1}\delta_0, \delta_0 \rangle$ for $z \in \rho(A)$.

Exercise 6.2
Give examples of positive operators A and B such that $A \leq B$ holds but $A^2 \leq B^2$ does not.

Exercise 6.3
Consider $\mathbb{C}^2$ as a Hilbert space with the standard inner product, and identify $\mathbf{B}(\mathbb{C}^2)$ with the matrix algebra $\mathbf{M}_2(\mathbb{C})$. Find the polar decomposition of the following matrices:

$$\begin{pmatrix} -1 & -2 \\ 2 & 1 \end{pmatrix}, \quad \begin{pmatrix} 1 & 0 \\ 1 & 0 \end{pmatrix}.$$

Exercise 6.4
Find the polar decomposition of the operator W_α in Problem 2.3.

Exercise 6.5
Let Ω be a compact Hausdorff space. For the Banach algebra $C(\Omega)$, show the following:

(1) $C(\Omega)^{-1} = \{f \in C(\Omega); \ 0 \notin f(\Omega)\}$.
(2) $\sigma(f) = f(\Omega)$ for $f \in C(\Omega)$.

Exercise 6.6
We call $A(\mathbb{D})$ in Exercise 1.5 the *disk algebra*. Through the restriction map $f \mapsto f|_{\partial\mathbb{D}}$, we can identify $A(\mathbb{D})$ with a closed subalgebra of $C(\partial\mathbb{D})$. Show that, in general, the spectrum of $f \in A(\mathbb{D})$ as an element in $A(\mathbb{D})$ is different from the spectrum of f as an element in $C(\partial\mathbb{D})$.

Chapter 7

Compact Operators on Banach Spaces

In this chapter, we first introduce compact operators, which are the most useful among bounded operators, and derive their basic properties. Next, we develop the theory of the Fredholm operators and derive the spectral properties of compact operators as an application. The index of a Fredholm operator is a typical example of an invariant obtained from infinite-dimensional spaces, and it plays a very important role in various fields in mathematics.

7.1   The Basic Properties of Compact Operators

Let us first review compactness of metric spaces. For a metric space (X, d), the following conditions are equivalent:

- X is compact.
- X is sequentially compact.
- X is totally bounded and complete.

When (X, d) is a metric space, we say that a subset Y of X is *totally bounded* if the following holds: $\forall \varepsilon > 0$, $\exists y_1, y_2, \ldots, y_n \in Y$, $Y \subset \bigcup_{i=1}^{n} B(y_i, \varepsilon)$. When Y is totally bounded, so is its closure. Thus, when (X, d) is complete, the subset Y is relatively compact if and only if it is totally bounded.

Theorem 7.1 (Ascoli–Arzelà theorem). *Let Ω be a compact Hausdorff space. Then, the following conditions on a subset K of $C(\Omega)$ are equivalent:*

(1) *K is relatively compact with respect to the topology given by $\|\cdot\|_\infty$.*
(2) *The following two conditions hold:*
   (a) *For every $\omega \in \Omega$, the set $\{f(\omega)\}_{f \in K}$ is bounded.*
   (b) *For every $\varepsilon > 0$ and every $\omega \in \Omega$, there exists a neighborhood U_ω of ω such that for every $f \in K$ and every $\xi \in U_\omega$, we have $|f(\omega) - f(\xi)| < \varepsilon$ (equicontinuity).*

Proof. (1) $\implies$ (2). Since $\overline{K}^{\|\cdot\|_\infty}$ is compact and the norm is a continuous function, the set $\{\|f\|_\infty\}_{f \in K}$ is bounded, and hence (a) holds.

Since K is totally bounded, for every $\varepsilon > 0$, there exist $f_1, f_2, \ldots, f_n \in K$ satisfying $K \subset \bigcup_{i=1}^n B(f_i, \frac{\varepsilon}{3})$. For $\omega \in \Omega$, let

$$U_\omega = \bigcap_{i=1}^n \left\{ \xi \in \Omega; \ |f_i(\omega) - f_i(\xi)| < \frac{\varepsilon}{3} \right\}.$$

Then, U_ω is a neighborhood of ω. For every $f \in K$, we choose $1 \le i \le n$ satisfying $\|f - f_i\|_\infty < \varepsilon/3$. For every $\xi \in U_\omega$, we have

$$|f(\omega) - f(\xi)| \le |f(\omega) - f_i(\omega)| + |f_i(\omega) - f_i(\xi)| + |f_i(\xi) - f(\xi)| < \varepsilon,$$

and (b) holds.

(2) $\implies$ (1). Let $\varepsilon > 0$. We take a neighborhood U_ω of each $\omega \in \Omega$ satisfying (b) for $\varepsilon/3$ in place of ε. Since $\bigcup_{\omega \in \Omega} U_\omega$ is an open cover of Ω and Ω is compact, there exist $\omega_1, \omega_2, \ldots, \omega_m \in \Omega$ satisfying $\bigcup_{j=1}^m U_{\omega_j} = \Omega$. We define a linear map $\Phi : C(\Omega) \to \mathbb{C}^m$ by

$$\Phi(f) = (f(\omega_1), f(\omega_2), \ldots, f(\omega_m)).$$

Then, (a) shows that $\Phi(K)$ is a bounded subset of $\mathbb{C}^m$. Since $\overline{\Phi(K)}$ is compact, it is totally bounded, and there exist $f_1, f_2, \ldots, f_n \in K$ satisfying $\Phi(K) \subset \bigcup_{i=1}^n B(\Phi(f_i), \varepsilon/3)$. Here, we assume that the metric of $\mathbb{C}^m$ is given by $\|\cdot\|_\infty$. Thus, for every $f \in K$, there exists $1 \le i \le n$

satisfying $\|\Phi(f) - \Phi(f_i)\|_\infty < \varepsilon/3$. For every $\omega \in \Omega$, we can choose $1 \le j \le m$ satisfying $\omega \in U_{\omega_j}$. Then,

$$|f(\omega) - f_i(\omega)| \le |f(\omega) - f(\omega_j)| + |f(\omega_j) - f_i(\omega_j)|$$
$$+ |f_i(\omega_j) - f_i(\omega)| < \varepsilon,$$

and we get $\|f - f_i\|_\infty < \varepsilon$. Thus, K is totally bounded. $\qquad\square$

Remark 7.1. Under condition (b), condition (a) is equivalent to $\{\|f\|_\infty\}_{f \in K}$ being bounded.

In the remainder of this section, we assume that X, Y, and Z are Banach spaces.

Definition 7.1. We say that $T \in \mathbf{B}(X, Y)$ is a *compact operator* if TB_X is relatively compact. This is equivalent to each of the following conditions:

(1) Whenever B is a bounded subset of X, its image TB is relatively compact.
(2) For every bounded sequence $\{x_n\}_{n=1}^\infty$ in X, there exists a convergent subsequence of $\{Tx_n\}_{n=1}^\infty$.

We denote by $\mathbf{K}(X, Y)$ the set of compact operators from X to Y and write $\mathbf{K}(X) = \mathbf{K}(X, X)$. The second condition shows that $\mathbf{K}(X, Y)$ is a subspace of $\mathbf{B}(X, Y)$.

Lemma 7.1. *Let $S \in \mathbf{B}(X, Y)$ and $T \in \mathbf{B}(Y, Z)$.*

(1) *If $S \in \mathbf{K}(X, Y)$, we have $TS \in \mathbf{K}(X, Z)$.*
(2) *If $T \in \mathbf{K}(Y, Z)$, we have $TS \in \mathbf{K}(X, Z)$.*

Proof. (1) Since SB_X is relatively compact and T is continuous, the image TSB_X is relatively compact.

(2) Since SB_X is bounded and $T \in \mathbf{K}(Y, Z)$, the image TSB_X is relatively compact. $\qquad\square$

Example 7.1. Let $X = L^1[0,1]$, let $Y = C[0,1]$, and let $k \in C[0,1]^2$. For $f \in X$, we set $A_k f(s) = \int_0^1 k(s,t) f(t) dt$. Then, $A_k f \in Y$. Since

$$|A_k f(s)| \le \int_0^1 |k(s,t)||f(t)| dt \le \|k\|_{C[0,1]^2} \int_0^1 |f(t)| dt,$$

we have $A_k \in \mathbf{B}(X,Y)$, and $\|A_k\| \le \|k\|_{C[0,1]^2}$ holds.

Using the Ascoli–Arzelà theorem, we show $A_k \in \mathbf{K}(X,Y)$. If $f \in B_X$, we have $\|A_k f\|_\infty \le \|k\|_{C[0,1]^2}$. Since k is uniformly continuous on $[0,1]^2$, for every $\varepsilon > 0$, there exists $\delta > 0$ such that if $(s,t),(s',t') \in [0,1]^2$ satisfy $|s - s'| + |t - t'| < \delta$, we get $|k(s,t) - k(s',t')| < \varepsilon$. In particular, if $|s - s'| < \delta$, we get

$$|A_k f(s) - A_k f(s')| \le \int_0^1 |k(s,t) - k(s',t)||f(t)| dt$$

$$\le \varepsilon \int_0^1 |f(t)| dt \le \varepsilon \|f\|_1,$$

and $A_k B_X$ is equicontinuous. Thus, $A_k B_X$ is relatively compact and $A_k \in \mathbf{K}(X,Y)$.

Note that for $1 \le p \le \infty$, the embedding maps $C[0,1] \hookrightarrow L^p[0,1] \hookrightarrow L^1[0,1]$ are continuous. Thus, if $1 \le p, q \le \infty$, the composition of A_k and either of the following maps is a compact operator:

$$C[0,1] \hookrightarrow L^p[0,1] \hookrightarrow L^1[0,1] \overset{A_k}{\to} C[0,1] \hookrightarrow L^q[0,1].$$

Lemma 7.2. *The space* $\mathbf{K}(X,Y)$ *is a closed subspace of* $\mathbf{B}(X,Y)$.

Proof. Assume that a sequence $\{T_n\}_{n=1}^\infty$ in $\mathbf{K}(X,Y)$ converges to $T \in \mathbf{B}(X,Y)$ in the operator norm. Then, we have the following: $\forall \varepsilon > 0, \exists N \in \mathbb{N}, \forall n \ge N, \|T_n - T\| < \varepsilon/3$. Since $T_N B_X$ is totally bounded, there exist $x_1, x_2, \ldots, x_m \in B_X$ satisfying $T_N B_X \subset \bigcup_{i=1}^m B(T_N x_i, \varepsilon/3)$. For every $x \in B_X$, there exists $1 \le i \le m$ satisfying $\|T_N x - T_N x_i\| < \varepsilon/3$, and

$$\|Tx - Tx_i\| \le \|Tx - T_N x\| + \|T_N x - T_N x_i\| + \|T_N x_i - Tx_i\| < \varepsilon.$$

Thus, $T B_X$ is totally bounded. $\qquad\qquad\square$

Corollary 7.1. *The space* $\mathbf{K}(X)$ *is a closed two-sided ideal of* $\mathbf{B}(X)$. *In particular,* $\mathbf{K}(X)$ *is a Banach algebra.*

Note that Theorem 1.7 shows the following: $I \in \mathbf{K}(X) \Longleftrightarrow \dim X < \infty$.

Corollary 7.2. $\mathbf{K}(X) \cap \mathbf{B}(X)^{-1} \neq \emptyset \Longleftrightarrow \dim X < \infty$.

Definition 7.2. We say that $T \in \mathbf{B}(X,Y)$ is a *finite rank operator* if $\dim \mathcal{R}(T) < \infty$. We denote by $\mathbf{F}(X,Y)$ the set of finite rank operators from X to Y and write $\mathbf{F}(X) = \mathbf{F}(X,X)$. For $T \in \mathbf{F}(X,Y)$, we define the rank of T by $\operatorname{rank} T = \dim \mathcal{R}(T)$.

When $T \in \mathbf{F}(X,Y)$, the image TB_X is a bounded set in the finite-dimensional space $\mathcal{R}(T)$, and it is relatively compact. Thus, $\mathbf{F}(X,Y)$ is a subspace of $\mathbf{K}(X,Y)$. Also, $\mathbf{F}(X)$ is a two-sided ideal of $\mathbf{K}(X)$ and $\mathbf{B}(X)$.

Theorem 7.2. *For* $T \in \mathbf{B}(X,Y)$, *we consider the following two conditions:*

(1) $T \in \mathbf{K}(X,Y)$.
(2) *Whenever a sequence* $\{x_n\}_{n=1}^{\infty}$ *in* X *weakly converges to* $x \in X$, *the sequence* $\{Tx_n\}_{n=1}^{\infty}$ *converges to* Tx *in norm.*

In general, the implication (1) $\Longrightarrow$ (2) *holds. If* X *is reflexive, the implication* (2) $\Longrightarrow$ (1) *holds.*

Proof. Let $T \in \mathbf{K}(X,Y)$, and assume that a sequence $\{x_n\}_{n=1}^{\infty}$ in X weakly converges to $x \in X$. Note that the principle of uniform boundedness shows that $\{x_n\}_{n=1}^{\infty}$ is bounded. Assume that (2) does not hold. Then, there would exist $\varepsilon > 0$ and a subsequence $\{x_{n_k}\}_{k=1}^{\infty}$ satisfying $\|Tx_{n_k} - Tx\| \geq \varepsilon$ for all $k \in \mathbb{N}$. Since $T \in \mathbf{K}(X,Y)$, the sequence $\{Tx_{n_k}\}_{k=1}^{\infty}$ has a norm convergent subsequence, and we may assume that $\{Tx_{n_k}\}_{k=1}^{\infty}$ converges in norm from the beginning. Let $y = \lim_{k \to \infty} Tx_{n_k}$. Since $\varphi \circ T \in X^*$ for every $\varphi \in Y^*$, we have $\lim_{k \to \infty} \varphi \circ T(x_{n_k}) = \varphi \circ T(x)$, and $\varphi(y) = \varphi(Tx)$. As this holds for every $\varphi \in Y^*$, we get $y = Tx$, but this contradicts $\|Tx_{n_k} - Tx\| \geq \varepsilon$. Thus, (2) holds.

Assume that X is reflexive and (2) holds. Let $\{x_n\}_{n=1}^{\infty}$ be a bounded sequence in X. By Corollary 5.6, there exists a weakly convergent subsequence $\{x_{n_k}\}_{k=1}^{\infty}$. From (2), the sequence $\{Tx_{n_k}\}_{k=1}^{\infty}$ converges in norm, and we get $T \in \mathbf{K}(X, Y)$. $\qquad\square$

We say that T satisfying condition (2) is a *completely continuous operator*.

For $T \in \mathbf{B}(X, Y)$, we define $T' \in \mathbf{B}(Y^*, X^*)$ by $T'\varphi = \varphi \circ T$.

Theorem 7.3 (Schauder theorem). *For $T \in \mathbf{B}(X, Y)$, the following two conditions are equivalent:*

(1) $T \in \mathbf{K}(X, Y)$.
(2) $T' \in \mathbf{K}(Y^*, X^*)$.

Proof. (1) $\Longrightarrow$ (2). Let $\Omega = \overline{TB_X}$. Then, Ω is compact. We denote by $\hat{\varphi} \in C(\Omega)$ the restriction of $\varphi \in Y^*$ to Ω. Since for $\varphi, \psi \in B_{Y^*}$,

$$\|T'\varphi - T'\psi\| = \sup_{x \in B_X} |\varphi(Tx) - \psi(Tx)|$$

$$= \sup_{y \in \Omega} |\hat{\varphi}(y) - \hat{\psi}(y)| = \|\hat{\varphi} - \hat{\psi}\|_{\infty}$$

holds, in order to show that $\{T'\varphi\}_{\varphi \in B_{Y^*}}$ is totally bounded, it suffices to show that $\{\hat{\varphi}\}_{\varphi \in B_{Y^*}}$ is totally bounded in $C(\Omega)$. For $\varphi \in B_{Y^*}$, we have $\|\hat{\varphi}\|_{\infty} \le \|T\|$. For $\varphi \in B_{Y^*}$ and $y_1, y_2 \in \Omega$, we have $|\hat{\varphi}(y_1) - \hat{\varphi}(y_2)| \le \|y_1 - y_2\|$, which shows that $\{\hat{\varphi}\}_{\varphi \in B_{Y^*}}$ is equicontinuous. Thus, it follows from Ascoli–Arzelà theorem that $\{\hat{\varphi}\}_{\varphi \in B_{Y^*}}$ is totally bounded.

(2) $\Longrightarrow$ (1). The above argument implies $T'' \in \mathbf{K}(X^{**}, Y^{**})$, and $T = T''|_X \in \mathbf{K}(X, Y)$. $\qquad\square$

7.2 Fredholm Operators#

In this section, we assume that X is a Banach space. For $T \in \mathbf{B}(X)$, we set $\operatorname{coker} T = X/\mathcal{R}(T)$.

Definition 7.3. We say that $T \in \mathbf{B}(X)$ is a *Fredholm operator* if $\dim \ker T < \infty$ and $\dim \operatorname{coker} T < \infty$.

When T is a Fredholm operator, we define its *index* by

$$\operatorname{ind} T = \dim \ker T - \dim \operatorname{coker} T.$$

We denote by $\mathrm{FR}(X)$ the set of Fredholm operators of X and by $\mathrm{FR}_n(X)$ the set of Fredholm operators of index n.

Example 7.2. If $\dim X < \infty$, we have $\mathbf{B}(X) = \mathrm{FR}(X) = \mathrm{FR}_0(X)$. The second equality follows from the dimension theorem.

Example 7.3. Let V be the unilateral shift of ℓ^2. Then, V is a Fredholm operator and

$$\mathrm{ind}\, V = \dim \ker V - \dim \mathrm{coker}\, V = 0 - 1 = -1.$$

Problem 7.1. Show $\mathbf{B}(X)^{-1} \subset \mathrm{FR}_0(X)$ and

$$\mathbf{B}(X)^{-1}\,\mathrm{FR}_n(X) = \mathrm{FR}_n(X)\mathbf{B}(X)^{-1} = \mathrm{FR}_n(X).$$

The purpose of this section is to show the following theorem.

Theorem 7.4. *Let X be a Banach space,*

(1) *The set $\mathrm{FR}(X)$ is open in $\mathbf{B}(X)$, and the map $\mathrm{ind} : \mathrm{FR}(X) \to \mathbb{Z}$ is continuous.* (continuity)
(2) *For $T \in \mathrm{FR}(X)$ and $K \in \mathbf{K}(X)$, we have $T + K \in \mathrm{FR}(X)$, and the equality $\mathrm{ind}(T + K) = \mathrm{ind}(T)$ holds.* (stability)
(3) *For $S, T \in \mathrm{FR}(X)$, we have $ST \in \mathrm{FR}(X)$, and the equality $\mathrm{ind}(ST) = \mathrm{ind}\, S + \mathrm{ind}\, T$ holds.* (additivity)

We prove the theorem in several steps. The following argument is a modification of a short proof in the case of Hilbert spaces given by Pedersen (1989). We discuss an application of the theorem to the spectra of compact operators after proving the theorem.

For a closed subspace M of X, if there exists a closed subspace N of X satisfying $M \cap N = \{0\}$ and $M + N = X$, we say that M is complemented, and we call N a complement of M. When this happens, Example 4.1 shows that the map $M \oplus_1 N \to X$, $(m, n) \mapsto m+n$ is a Banach space isomorphism. In general, M is not necessarily complemented, and even when it is the case, a complement is not unique.

Problem 7.2. Show the following for a closed subspace M of X.

(1) If $\dim M < \infty$, then M is complemented.
(2) If $\dim X/M < \infty$, then M is complemented.

Lemma 7.3. *For a Fredholm operator $T \in \mathrm{FR}(X)$, its range $\mathcal{R}(T)$ is closed.*

Proof. Since $\dim \ker T < \infty$, we choose a complement M of $\ker T$. Since $\dim X/\mathcal{R}(T) < \infty$, we choose a basis $\{a_i\}_{i=1}^n$ of $X/\mathcal{R}(T)$, and choose $x_i \in X$, with $x_i + \mathcal{R}(T) = a_i$. Let $N = \mathrm{span}\{x_i\}_{i=1}^n$. Then, N is finite-dimensional and hence closed. We have $\mathcal{R}(T) \cap N = \{0\}$ and $\mathcal{R}(T) + N = X$. We define a map $T_1 : M \oplus_1 N \to X$ by $(m,n) \mapsto Tm + n$. Since

$$\|T_1(m,n)\| \leq \|Tm\| + \|n\| \leq \max\{\|T\|, 1\}\|(m,n)\|_1,$$

we have $T_1 \in \mathbf{B}(M \oplus_1 N, X)$, and T_1 is a bijection. Thus, the Banach inverse mapping theorem shows that T_1^{-1} is also continuous. Since $M \oplus_1 \{0\}$ is closed in $M \oplus_1 N$, so is $T_1(M \oplus_1 \{0\}) = \mathcal{R}(T)$ in X. $\square$

Lemma 7.4. *For $\lambda \in \mathbb{C}\backslash\{0\}$ and $K \in \mathbf{K}(X)$, we have $\lambda I - K \in \mathrm{FR}(X)$.*

Proof. Since $KB_{\ker(\lambda I - K)} = \lambda B_{\ker(\lambda I - K)}$ and $K \in \mathbf{K}(X)$, we have $\dim \ker(\lambda I - K) < \infty$.

Let $T = \lambda I - K$. We choose a complement M of $\ker T$, and let $T_1 = T|_M : M \to X$ be the restriction of T to M. Then, T_1 is bounded below. Indeed, if it were not bounded below, there would exist a sequence $\{x_n\}_{n=1}^\infty$ of M satisfying $\|x_n\| = 1$ and $\|T_1 x_n\| \to 0$ as $n \to \infty$. Since $K \in \mathbf{K}(X)$, there exists a subsequence $\{x_{n_k}\}_{k=1}^\infty$ such that $\{Kx_{n_k}\}_{k=1}^\infty$ is convergent. Since $\{\lambda x_{n_k} - Kx_k\}_{k=1}^\infty$ converges to 0 and $\lambda \neq 0$, the sequence $\{x_{n_k}\}_{k=1}^\infty$ is convergent too. Denoting its limit by x, we get $x \in M$, $\|x\| = 1$, and $x \in \ker(\lambda I - K)$, which is a contradiction. Thus, T_1 is bounded below.

Since T_1 is bounded below, Lemma 6.5 shows that $\mathcal{R}(T) = \mathcal{R}(T_1)$ is closed. From Theorem 3.3(2), we have

$$(X/\mathcal{R}(T))^* = \mathcal{R}(T)^\perp = \ker T' = \ker(\lambda I_{X^*} - K').$$

Since the Schauder theorem shows that $K' \in \mathbf{K}(X^*)$, we have $\dim(X/\mathcal{R}(T))^* < \infty$, and $\dim(X/\mathcal{R}(T)) < \infty$. $\square$

Lemma 7.5. *For every* $T \in \mathrm{FR}(X)$, *the following hold*:

(1) *There exists* $S \in FR(X)$ *satisfying* $I - ST, I - TS \in \mathbf{F}(X)$.
(2) *If* $T \in \mathrm{FR}_0(X)$, *there exists* $R \in \mathbf{F}(X)$ *satisfying* $T + R \in \mathbf{B}(X)^{-1}$.

Proof. We choose a complement M of $\ker T$ and a complement N of $\mathcal{R}(T)$. Then,

$$X \cong \ker T \oplus_1 M \cong N \oplus_1 \mathcal{R}(T).$$

(1) Let $T_1 \in \mathbf{B}(M, \mathcal{R}(T))$ be the restriction of T to M. Then, the Banach inverse mapping theorem shows that T_1 is invertible. For $n \in N$ and $y \in \mathcal{R}(T)$, we set $S(n + y) = T_1^{-1}y$. Then, $S \in \mathrm{FR}(X)$ and $TS(n + y) = y$ hold. Also, for $x \in \ker T$ and $m \in M$, we have $ST(x + m) = m$. Thus, $I - ST, I - TS \in \mathbf{F}(X)$.

(2) If $\mathrm{ind}\, T = 0$, we have $\dim \ker T = \dim \mathrm{coker}\, T = \dim N$. Since $\ker T$ and N are finite-dimensional spaces of the same dimension, there exists an isomorphism $R_0 : \ker T \to N$. For $x \in \ker T$ and $m \in M$, we set $R(x + m) = R_0 x$. Then, $R \in \mathbf{F}(X)$, and the Banach inverse mapping theorem shows $R + T \in \mathbf{B}(X)^{-1}$. $\qquad\square$

Problem 7.3. Let $\mathcal{I}$ be a two-sided closed ideal of a Banach algebra $\mathcal{A}$. Show that $\mathcal{A}/\mathcal{I}$ is a Banach algebra with respect to the quotient norm.

Recall that $\mathbf{K}(X)$ is a closed two-sided ideal of $\mathbf{B}(X)$. We call $\mathbf{Q}(X) = \mathbf{B}(X)/\mathbf{K}(X)$ the *Calkin algebra*. The Calkin algebra $\mathbf{Q}(X)$ is a Banach algebra. Let $\pi : \mathbf{B}(X) \to \mathbf{Q}(X)$ be the quotient map.

Theorem 7.5 (Atkinson theorem). *For* $T \in \mathbf{B}(X)$, *the following conditions are equivalent*:

(1) $T \in \mathrm{FR}(X)$.
(2) $\pi(T) \in \mathbf{Q}(X)^{-1}$.

Proof. (1) $\Longrightarrow$ (2). Since there exists $S \in \mathbf{B}(X)$ satisfying $I - ST, I - TS \in \mathbf{F}(X)$, we get $\pi(S) = \pi(T)^{-1}$.

(2) $\implies$ (1). We choose $S \in \mathbf{B}(X)$ satisfying $\pi(S) = \pi(T)^{-1}$, and set $K_1 = I - ST$, $K_2 = I - TS$. Then, $K_1, K_2 \in \mathbf{K}(X)$, and

$$\dim \ker T \leq \dim \ker(ST) = \dim \ker(I - K_1) < \infty,$$
$$\dim \operatorname{coker} T \leq \dim \operatorname{coker}(TS) = \dim \operatorname{coker}(I - K_2) < \infty.$$

Thus, $T \in \mathrm{FR}(X)$. $\qquad\square$

Note that $\mathbf{Q}(X)^{-1}$ is an open subset of $\mathbf{Q}(X)$ and is a group. The continuity of the quotient map $\pi : \mathbf{B}(X) \to \mathbf{Q}(X)$ shows the following.

Corollary 7.3. *The set* $\mathrm{FR}(X)$ *is open in* $\mathbf{B}(X)$, *and* $\mathrm{FR}(X)\,\mathrm{FR}(X) \subset \mathrm{FR}(X)$ *holds.*

For two Banach space X_1, X_2 and $T_i \in \mathbf{B}(X_i)$, $i = 1, 2$, we define $T_1 \oplus T_2 \in \mathbf{B}(X_1 \oplus_1 X_2)$ by $(T_1 \oplus T_2)(x_1, x_2) = (T_1 x_1, T_2 x_2)$. Then, $T_1 \oplus T_2 \in \mathrm{FR}(X_1 \oplus_1 X_2) \iff T_1 \in \mathrm{FR}(X_1)$ and $T_2 \in \mathrm{FR}(X_2)$, and when this is the case, we have $\operatorname{ind}(T_1 \oplus T_2) = \operatorname{ind} T_1 + \operatorname{ind} T_2$.

Lemma 7.6. *We have* $\mathrm{FR}_0(X) = \mathbf{B}(X)^{-1} + \mathbf{F}(X)$.

Proof. Since we have already shown that $\mathrm{FR}_0(X) \subset \mathbf{B}(X)^{-1} + \mathbf{F}(X)$ in Lemma 7.5(2), we show the converse inclusion. Although we should show $W + F \in \mathrm{FR}_0(X)$ for every $W \in \mathbf{B}(X)^{-1}$ and every $F \in \mathbf{F}(X)$, since $W + F = W(I + W^{-1}F)$ and $W^{-1}F \in \mathbf{F}(X)$, it suffices to show the statement for $W = I$.

We choose a basis $\{y_i\}_{i=1}^n$ of $\mathcal{R}(F)$ and take its dual basis $\{y_i^*\}_{i=1}^n \subset \mathcal{R}(F)^*$. Taking $x_i \in X$ satisfying $y_i = Fx_i$ and setting $\varphi_i = y_i^* \circ F \in X^*$, we get $\varphi_j(x_i) = \delta_{i,j}$. Since $\ker F = \bigcap_{i=1}^n \ker \varphi_i$, the space $M = \operatorname{span}\{x_i\}_{i=1}^n$ is a complement of $\ker F$. We decompose y_i as $y_i = a_i + m_i$, $a_i \in \ker F$, $m_i \in M$, and let $N_0 = \operatorname{span}\{a_i\}_{i=1}^n$. Then, since N_0 is a finite-dimensional subspace of $\ker F$, there exists a complement $N_1 \subset \ker F$ of N_0. Let $M_1 = M + N_0$, which is finite-dimensional. Then, we have $X = M_1 + N_1$, and $M_1 \cap N_1 = \{0\}$. By construction, we have $N_1 \subset \ker F$ and $\mathcal{R}(F) \subset M_1$. Thus, we get

$$\operatorname{ind}(I + F) = \operatorname{ind}(I + F)|_{M_1} + \operatorname{ind}(I + F)|_{N_1} = 0 + \operatorname{ind} I_{N_1} = 0.$$

$\qquad\square$

Since $\mathbf{B}(X)^{-1}$ is an open subset of $\mathbf{B}(X)$ and is a group, the following holds.

Corollary 7.4. *The set* $\mathrm{FR}_0(X)$ *is open in* $\mathbf{B}(X)$, *and* $\mathrm{FR}_0(X)\,\mathrm{FR}_0(X) \subset \mathrm{FR}_0(X)$ *holds.*

Since we use the assertion of the following problem in the proof of Theorem 7.4, solve it by using only the definition of the index.

Problem 7.4. Let V be the unilateral shift of ℓ^2. For $n \in \mathbb{Z}$, we set

$$V^{(n)} = \begin{cases} V^n, & n > 0, \\ I, & n = 0, \\ V^{*-n}, & n < 0. \end{cases}$$

Then, show that $\mathrm{ind}(V^{(m)}V^{(n)}) = -(m+n)$ holds for any $m, n \in \mathbb{Z}$.

Proof of Theorem 7.4. (1) We have already seen that $\mathrm{FR}(X)$ is an open subset of $\mathbf{B}(X)$. To show the continuity of the index, it suffices to show that $\mathrm{FR}_n(X)$ is open. For $T \in \mathrm{FR}_n(X)$, we have $T \oplus V^{(n)} \in \mathrm{FR}_0(X \oplus_1 \ell^2)$. As $\mathrm{FR}_0(X \oplus_1 \ell^2)$ is an open subset of $\mathbf{B}(X \oplus_1 \ell^2)$, there exists $\varepsilon > 0$ such that if $R \in \mathbf{B}(X \oplus_1 \ell^2)$ satisfies $\|R - (T \oplus V^{(n)})\| < \varepsilon$, we have $R \in \mathrm{FR}_0(X \oplus_1 \ell^2)$. In particular, if $S \in \mathbf{B}(X)$ satisfies $\|S - T\| < \varepsilon$, we get $S \oplus V^{(n)} \in \mathrm{FR}_0(X \oplus_1 \ell^2)$, and $S \in \mathrm{FR}_n(X)$. Thus, $\mathrm{FR}_n(X)$ is an open set.

(2) Let $T \in \mathrm{FR}(X)$, and let $K \in \mathbf{K}(X)$. By the Atkinson theorem, we have $T + K \in \mathrm{FR}(X)$. Since the map $[0,1] \ni t \mapsto T + tK \in \mathrm{FR}(X)$ is continuous, the continuity of the index implies $\mathrm{ind}(T + K) = \mathrm{ind}\, T$.

(3) Let $S \in \mathrm{FR}_m(X)$, and let $T \in \mathrm{FR}_n(X)$. Then, $ST \in \mathrm{FR}(X)$ follows from the Atkinson theorem. Let $S' = S \oplus V^{(m)}$, and let $T' = T \oplus V^{(n)}$. Then, $S', T' \in \mathrm{FR}_0(X \oplus_1 \ell^2)$, and we have $S'T' = ST \oplus V^{(m)}V^{(n)} \in \mathrm{FR}_0(X \oplus_1 \ell^2)$. Since $\mathrm{ind}(V^{(m)}V^{(n)}) = -(m+n)$, we get $\mathrm{ind}(ST) = m + n$. $\qquad\square$

Let $K \in \mathbf{K}(X)$, and let $\lambda \in \mathbb{C}\backslash\{0\}$. From the fact that $\lambda I - K$ is a Fredholm operator of index 0, we can obtain the following theorem, called the Fredholm alternative, or the Riesz–Schauder theorem.

Corollary 7.5. *If* $K \in \mathbf{K}(X)$, *we have* $\sigma(K)\backslash\{0\} = \sigma_{\mathrm{p}}(K)\backslash\{0\}$. *For* $\lambda \in \sigma_{\mathrm{p}}(K)\backslash\{0\}$, *we have* $\dim\ker(\lambda I - K) = \dim\mathrm{coker}(\lambda I - K) < \infty$.

Proof. If $\lambda \in \mathbb{C}\backslash\{0\}$ is not an eigenvalue of K, we have $\ker(\lambda I - K) = \{0\}$, and $\lambda I - T$ is a surjection because $\operatorname{ind}(\lambda I - T) = 0$. Thus, the Banach inverse mapping theorem shows that $\lambda I - T$ is invertible and $\lambda \notin \sigma(K)$. $\qquad\square$

Recall that if $\dim X = \infty$, a compact operator $K \in \mathbf{K}(X)$ is not invertible, and $0 \in \sigma(K)$.

Theorem 7.6. *For* $K \in \mathbf{K}(X)$, *the spectrum* $\sigma(K)$ *is an at most countable set, and the only accumulation point is 0 if it exists.*

Proof. We assume that $\lambda \in \sigma(K)\backslash\{0\}$ is an accumulation point of $\sigma(K)$ and deduce a contradiction. We choose a sequence $\{\lambda_n\}_{n=1}^{\infty}$ in $\sigma(K)\backslash\{0\}$ converging to λ and satisfying $\lambda_m \neq \lambda_n$ for any $m \neq n$. Let $L = \inf_{n\in\mathbb{N}} |\lambda_n|$, which satisfies $L > 0$. Since $\lambda_n \in \sigma_{\mathrm{p}}(K)$, there exists $x_n \in \ker(\lambda_n I - K)$, with $\|x_n\| = 1$. Since $\{\lambda_n\}_{n=1}^{\infty}$ are distinct, the set $\{x_n\}_{n=1}^{\infty}$ is linearly independent, and $X_n = \operatorname{span}\{x_k\}_{k=1}^{n}$, $n \in \mathbb{N}$, satisfy $X_{n-1} \subsetneq X_n$. Thus, by Lemma 1.1, there exist $y_n \in X_n$ such that $\|y_n\| = 1$ and $d(y_n, X_{n-1}) \geq 1/2$. Note that we have $KX_n \subset X_n$. Let $n < m$. Since y_m is uniquely decomposed as $y_m = z_{m-1} + c_m x_m$, $z_{m-1} \in X_{m-1}$, $c_m \in \mathbb{C}$, we get

$$Ky_m = Kz_{m-1} + c_m\lambda_m x_m = Kz_{m-1} - \lambda_m z_{m-1} + \lambda_m y_m,$$

and

$$\|Ky_n - Ky_m\| = \|Ky_n - Kz_{m-1} + \lambda_m z_{m-1} - \lambda_m y_m\| \geq \frac{|\lambda_m|}{2} \geq \frac{L}{2}.$$

This shows that $\{Ky_n\}_{n=1}^{\infty}$ has no convergent subsequence, which contradicts $K \in \mathbf{K}(X)$. Therefore, the only accumulation point of $\sigma(K)$ is 0 if it exists.

The above argument shows that

$$\left\{ \lambda \in \sigma(K); \ |\lambda| \geq \frac{1}{n} \right\}$$

is a compact set without accumulation points for all $n \in \mathbb{N}$, and it is a finite set. Therefore, $\sigma(K)$ is an at most countable set. $\qquad\square$

Exercises

Exercise 7.1
Show that the Hilbert cube

$$C = \left\{ (a_n)_{n=1}^{\infty} \in \ell^2; \ \forall n \in \mathbb{N}, \ |a_n| \leq \frac{1}{n} \right\}$$

is compact.

Exercise 7.2
The set of continuous functions on $[0, \infty)$ vanishing at infinity $C_0[0, \infty)$ is a Banach space with the norm $\|f\|_\infty = \sup_{t \in [0,\infty)} |f(t)|$.

(1) For $f \in L^2[0, \infty)$, we set $Tf(s) = (1 + s)^{-1} \int_0^s f(t)dt$. Show $|Tf(s)| \leq (1 + s)^{-1} \sqrt{s} \|f\|_2$.
(2) Show $T \in \mathbf{K}(L^2[0, \infty), C_0[0, \infty))$.

Exercise 7.3
We use the notation in Example 2.8 and Exercise 2.1 in the following problems:

(1) Show $[M_f, P_+] = M_f P_+ - P_+ M_f \in \mathbf{K}(L^2(\mathbb{T}))$ for $f \in C(\mathbb{T})$.
(2) Show $T_{fg} - T_f T_g \in \mathbf{K}(H^2(\mathbb{T}))$ for $f \in C(\mathbb{T})$ and $g \in L^\infty(\mathbb{T})$.
(3) Show $T_f \in \mathrm{FR}(H^2(\mathbb{T}))$ for $f \in C(\mathbb{T})^{-1}$.

Exercise 7.4
Show $T_f \notin \mathrm{FR}(H^2(\mathbb{T}))$ for $f \in C(\mathbb{T}) \backslash C(\mathbb{T})^{-1}$ in the following steps. First, note that there exists $0 \leq t_0 < 2\pi$, with $f(t_0) = 0$.

(1) For $0 < r < 1$, we define $h_r \in H^2(\mathbb{T})$ by

$$h_r(t) = \sqrt{1 - r^2} \sum_{n=0}^{\infty} r^n e^{in(t-t_0)} = \frac{\sqrt{1 - r^2}}{1 - re^{i(t-t_0)}}.$$

Show $\|h_r\|_2 = 1$ and w-$\lim_{r \to 1-0} h_r = 0$.
(2) Show $\lim_{r \to 1-0} \|T_f h_r\|_2 = 0$.
(3) Assume $T_f \in \mathrm{FR}(H^2(\mathbb{T}))$, and deduce a contradiction.

Exercise 7.5

Let $f \in C(\mathbb{T})^{-1}$, and let $w(f)$ be its winding number. Show the index formula $\text{ind}\, T_f = -w(f)$ in the following steps:

(1) Show that there exists a continuous function $\theta : \mathbb{R} \to \mathbb{R}$ satisfying $f(t) = |f(t)|e^{i\theta(t)}$. Show that the function θ is unique up to adding a number in $2\pi\mathbb{Z}$.

(2) We define the winding number of f by $w(f) = (\theta(2\pi) - \theta(0))/(2\pi) \in \mathbb{Z}$. Show that f can be continuously deformed to $e_{w(f)}(t) = e^{iw(f)t}$ in $C(\mathbb{T})^{-1}$.

(3) Show $\text{ind}\, T_{e_n} = -n$ for every $n \in \mathbb{Z}$.

(4) Show $\text{ind}\, T_f = -w(f)$.

Exercise 7.6

Let $\pi : \mathbf{B}(X) \to \mathbf{B}(X)/\mathbf{K}(X) = \mathbf{Q}(X)$ be the quotient map. For $T \in \mathbf{B}(X)$, we denote $\sigma(\pi(T))$ by $\sigma_e(T)$, and call it the *essential spectrum* of T. If $\lambda I - T$ is invertible, so is $\pi(\lambda I - T)$, and we have $\sigma_e(T) \subset \sigma(T)$. We call $\sigma_w(T) = \bigcap_{K \in \mathbf{K}(X)} \sigma(T + K)$ the *Weyl spectrum* of T. From $\pi(T + K) = \pi(T)$, we have $\sigma_e(T) \subset \sigma_w(T)$ in general.

(1) For the unilateral shift V of ℓ^2, show $\sigma_e(V) = S^1$.

(2) Show $\sigma_w(V) = \overline{\mathbb{D}}$.

Chapter 8

Detailed Accounts on Compact Operators on Hilbert Spaces

In this chapter, we develop a detailed analysis of compact operators on Hilbert spaces. A compact self-adjoint operator has a complete orthonormal system consisting of eigenvectors, which together with the polar decomposition of an operator leads to the Schmidt expansion of a general compact operator. The Hilbert–Schmidt class and the trace class are important classes of compact operators for applications.

8.1   Compact Self-Adjoint Operators

In this section, we assume that $\mathcal{H}$ is a Hilbert space, and we use the notation

$$\mathbf{K}(\mathcal{H})_{\mathrm{sa}} = \mathbf{K}(\mathcal{H}) \cap \mathbf{B}(\mathcal{H})_{\mathrm{sa}}, \quad \mathbf{K}(\mathcal{H})_+ = \mathbf{K}(\mathcal{H}) \cap \mathbf{B}(\mathcal{H})_+.$$

When $T \in \mathbf{K}(\mathcal{H})$ is a normal operator, we can show Corollary 7.5 and Theorem 7.6 more simply and directly, which we do first.

Proof of Corollary 7.5.   Since T is normal, we have $\sigma(T) = \sigma_{\mathrm{ap}}(T)$, and for every $\lambda \in \sigma(T) \backslash \{0\}$, there exists an approximate eigenvector $\{x_n\}_{n=1}^{\infty}$. Since T is a compact operator, we may and do assume that $\{Tx_n\}_{n=1}^{\infty}$ converges by taking a subsequence if necessary. As $\{\lambda x_n - Tx_n\}_{n=1}^{\infty}$ converges to zero, the sequence $\{x_n\}_{n=1}^{\infty}$ also converges, and we denote its limit by x. Then, $x \in \ker(\lambda I - T) \backslash \{0\}$.

Thus, we get $\sigma(T)\backslash\{0\} = \sigma_{\mathrm{p}}(T)\backslash\{0\}$. For $\lambda \in \sigma_{\mathrm{p}}(T)\backslash\{0\}$, the image $TB_{\ker(\lambda I - T)} = \lambda B_{\ker(\lambda I - T)}$ is relatively compact, and $\dim \ker(\lambda I - T) < \infty$. $\qquad\square$

Proof of Theorem 7.6. Let $\{\lambda_n\}_{n=1}^{\infty} \subset \sigma(T)\backslash\{0\}$, and choose $e_n \in \ker(\lambda_n I - T)$, with $\|e_n\| = 1$. Lemma 6.7(2) shows that $\{e_n\}_{n=1}^{\infty}$ is an ONS, and the Bessel inequality implies that it weakly converges to 0. The complete continuity of a compact operator shows that $\{Te_n\}_{n=1}^{\infty}$ converges to 0, and $\|Te_n\| = |\lambda_n| \to 0$ as $(n \to \infty)$. Thus, $\sigma(T)$ has the only accumulation point 0 if it exists. $\qquad\square$

Putting Theorem 6.11, Corollary 7.5, and Theorem 7.6 together, we can obtain the following spectral decomposition theorem for compact self-adjoint operators.

Theorem 8.1 (Spectral decomposition). *Let $A \in \mathbf{K}(\mathcal{H})_{\mathrm{sa}}$. Then, $\sigma(A)\backslash\{0\}$ is an at most countable set $\{\lambda_n\}_{=1}^{N} \subset \mathbb{R}$, $N \in \mathbb{N}_0 \cup \{\infty\}$, and we may arrange it so that $\{|\lambda_n|\}_{n=1}^{N}$ is monotone decreasing. If $N = \infty$, it converges to zero. Let P_n be the projection onto the eigenspace $\ker(\lambda_n I - A)$. Then, $P_n \in \mathbf{F}(\mathcal{H})$, and*

$$A = \sum_{n=1}^{N} \lambda_n P_n$$

converges in norm.

Corollary 8.1. *If $A \in \mathbf{K}(\mathcal{H})_{\mathrm{sa}}$ satisfies $\sigma(A) \cap (0, \infty) \neq \emptyset$, then*

$$\sup_{\|x\|=1} \langle Ax, x \rangle = \max_{\|x\|=1} \langle Ax, x \rangle = \text{ the largest eigenvalue of } A$$

holds.
For $T \in \mathbf{K}(\mathcal{H})$, we have

$$\|T\| = \max_{\|x\|=1} \|Tx\|,$$

which coincides with the square root of the largest eigenvalue of T^*T.

Proof. Since $\sigma(A)\backslash\{0\} = \sigma_{\mathrm{p}}(A)\backslash\{0\}$ and $\sigma(A) \cap (0, \infty) \neq \emptyset$, the largest eigenvalue M of A exists. Let $x \in \mathcal{H}$, with $\|x\| = 1$.

The Bessel inequality shows

$$\langle Ax, x \rangle = \sum_{n=1}^{N} \lambda_n \langle P_n x, x \rangle \leq M \sum_{n=1}^{N} \langle P_n x, x \rangle \leq M\|x\|^2 = M,$$

and the equality holds if x is an eigenvector associated with M. The second assertion follows from $\|Tx\|^2 = \langle T^*Tx, x \rangle$. $\qquad\square$

Definition 8.1. We say that $T \in \mathbf{B}(\mathcal{H})$ is a *diagonal operator* if there exists a CONS of $\mathcal{H}$ consisting of the eigenvectors of $T \in \mathbf{B}(\mathcal{H})$.

Corollary 8.2. *Every* $A \in \mathbf{K}(\mathcal{H})_{\mathrm{sa}}$ *is a diagonal operator.*

Proof. We have $\mathcal{H} = \ker A \oplus \overline{\mathcal{R}(A)}$, and the spectral decomposition shows

$$\overline{\mathcal{R}(A)} = \mathrm{span} \bigcup_{n=1}^{N} \ker(\lambda_n I - A).$$

Thus, if we choose a CONS for each eigenspace of A, their union is a CONS for $\mathcal{H}$. $\qquad\square$

Proposition 8.1. *Let* $T \in \mathbf{B}(\mathcal{H})$ *be a diagonal operator with a CONS* $\{e_i\}_{i \in I}$ *for* $\mathcal{H}$ *satisfying* $Te_i = \lambda_i e_i$, $\lambda_i \in \mathbb{C}$. *Then, the following conditions are equivalent:*

(1) $T \in \mathbf{K}(\mathcal{H})$.
(2) $\{\lambda_i\}_{i \in I}$ *converges to 0 at infinity, that is,* $\forall \varepsilon > 0$, $\exists F \Subset I$, $\forall i \in I \backslash F$, $|\lambda_i| < \varepsilon$.

Proof. (1) $\Longrightarrow$ (2). Assume that (2) does not hold. Then, there exists $\varepsilon > 0$ and a countable subset $\{i_n\}_{n=1}^{\infty} \subset I$ satisfying $|\lambda_n| \geq \varepsilon$ for all $n \in \mathbb{N}$. Then, $\{e_{i_n}\}_{n=1}^{\infty}$ weakly converges to 0, but $\|Te_{i_n}\| = |\lambda_{i_n}|$ does not converge to 0. Thus, T is not a compact operator.

(2) $\Longrightarrow$ (1). For $n \in \mathbb{N}$, we set $F_n = \{i \in I; |\lambda_i| \geq 1/n\}$. Then, F_n is a finite subset. Let E_n be the projection onto $\mathrm{span}\{e_i\}_{i \in F_n}$.

Then, $E_n \in \mathbf{F}(\mathcal{H})$. For every $x \in \mathcal{H}$, we have

$$\|(T - TE_n)x\|^2 = \|T \sum_{i \in I \setminus F_n} \langle x, e_i \rangle e_i\|^2 = \sum_{i \in I \setminus F_n} |\lambda_i|^2 |\langle x, e_i \rangle|^2$$

$$\leq \frac{1}{n^2} \sum_{i \in I \setminus F_n} |\langle x, e_i \rangle|^2 \leq \frac{1}{n^2} \|x\|^2,$$

and $\|T - TE_n\| \leq 1/n$, which shows that $\{TE_n\}_{n=1}^{\infty}$ converges to T in the operator norm. Since $TE_n \in \mathbf{F}(\mathcal{H}) \subset \mathbf{K}(\mathcal{H})$, we get $T \in \mathbf{K}(\mathcal{H})$. $\square$

Lemma 8.1. *For $T \in \mathbf{B}(\mathcal{H})$, the following holds:*

$$T \in \mathbf{K}(\mathcal{H}) \iff |T| \in \mathbf{K}(\mathcal{H}) \iff T^* \in \mathbf{K}(\mathcal{H}).$$

Proof. Let $T = W|T|$ be the polar decomposition. The statement follows from $|T| = W^*T = T^*W$, $T^* = |T|W^*$. $\square$

Corollary 8.3. $\mathbf{K}(\mathcal{H}) = \overline{\mathbf{F}(\mathcal{H})}^{\|\cdot\|}$.

Proof. We already know that $\mathbf{K}(\mathcal{H}) \supset \overline{\mathbf{F}(\mathcal{H})}^{\|\cdot\|}$, and we show the converse inclusion. Let $T \in \mathbf{K}(\mathcal{H})$, and let $T = W|T|$ be its polar decomposition. Since $|T| \in \mathbf{K}(\mathcal{H})_+$, we have the spectral decomposition $|T| = \sum_{n=1}^{N} \lambda_n P_n$. If $N < \infty$, we have $|T| \in \mathbf{F}(\mathcal{H})$ and $T \in \mathbf{F}(\mathcal{H})$. If $N = \infty$, for every $m \in \mathbb{N}$, we have

$$\left\| T - W \sum_{n=1}^{m} \lambda_n P_n \right\| \leq \|W\| \left\| |T| - \sum_{n=1}^{m} \lambda_n P_n \right\| = \lambda_{m+1}.$$

Since $\{\lambda_m\}_{m=1}^{\infty}$ converges to 0, we get $T \in \overline{\mathbf{F}(\mathcal{H})}^{\|\cdot\|}$. $\square$

Definition 8.2. For $x, y \in \mathcal{H}$, we define $x \otimes y^* \in \mathbf{F}(\mathcal{H})$ by $(x \otimes y^*)z = \langle z, y \rangle x$, and call it the *Schatten form*.

From the definition, the following hold:

(1) $(x \otimes y^*)(z \otimes w^*) = \langle z, y \rangle x \otimes w^*$,
(2) $(x \otimes y^*)^* = y \otimes x^*$,
(3) $\|x \otimes y^*\| = \|x\| \|y\|$,

(4) If $\mathcal{K}$ is a finite-dimensional subspace of $\mathcal{H}$ and $\{e_i\}_{i=1}^n$ is a CONS for $\mathcal{K}$, the operator $\sum_{i=1}^n e_i \otimes e_i^*$ is the projection onto $\mathcal{K}$.

Definition 8.3. Let $T = W|T|$ be the polar decomposition of $T \in \mathbf{K}(\mathcal{H})$. We denote by $\{s_n(T)\}_{n=1}^N$, $N \in \mathbb{N}_0 \cup \{\infty\}$, the set of eigenvalues of $|T|$, counting the multiplicities, in decreasing order and call it the *singular values* of T. There exists a CONS $\{e_n\}_{n=1}^N$ for $\ker T^\perp = \ker |T|^\perp = \ker W^\perp$ satisfying $|T|e_n = s_n(T)e_n$. Let $f_n = We_n$. Then, $\{f_n\}_{n=1}^N$ is a CONS for $\overline{\mathcal{R}(T)} = \mathcal{R}(W)$, and

$$T = \sum_{n=1}^N s_n(T)f_n \otimes e_n^*$$

converges in the operator norm. We call it the *Schmidt expansion* of T.

Remark 8.1. The Schmidt expansion is not unique. This is because there is ambiguity in the choice of $\{e_n\}_{n=1}^N$.

Corollary 8.1 shows $s_1(T) = \|T\|$. From

$$|T^*|^2 = TT^* = \sum_{n=1}^N s_n(T)^2 f_n \otimes f_n^*,$$

we have $s_n(T) = s_n(T^*)$.

In what follows, we set $s_n(T) = 0$ for $n > N$ if $N < \infty$, and we proceed with our discussion regardless of whether N is finite or not if there is no possibility of confusion.

The following characterization is very useful to derive detailed properties of the singular values.

Theorem 8.2. *Let $T \in \mathbf{K}(\mathcal{H})$, and let $n \in \mathbb{N}$. Then, the following holds:*

$$s_n(T) = \min\{\|T - F\|;\ F \in \mathbf{F}(\mathcal{H}),\ \operatorname{rank} F \leq n - 1\}.$$

Proof. Let $T = \sum_{n=1}^N s_n(T)f_n \otimes e_n^*$ be the Schmidt expansion of T. We show the statement for $n \leq N$. Assume that the rank of $F \in \mathbf{F}(\mathcal{H})$ is less than or equal to $n - 1$, and let $\mathcal{K} = \operatorname{span}\{e_k\}_{k=1}^n$. Then, there exists $x \in \ker F \cap \mathcal{K}$, with $\|x\| = 1$. Indeed, since the rank of $F|_\mathcal{K} : \mathcal{K} \to \mathcal{H}$ is less than or equal to $n-1$ and $\dim \mathcal{K} = n$, we have $\ker F|_\mathcal{K} \neq \{0\}$. Since

$$\|(T - F)x\|^2 = \left\| \sum_{k=1}^{n} s_k(T)\langle x, e_k \rangle f_k \right\|^2 = \sum_{k=1}^{n} s_k(T)^2 |\langle x, e_k \rangle|^2$$

$$\geq s_n(T)^2 \sum_{k=1}^{n} |\langle x, e_k \rangle|^2 = s_n(T)^2 \|x\|^2,$$

we have $\|T - F\| \geq s_n(T)$.

On the other hand, let $T_{n-1} = \sum_{k=1}^{n-1} s_k(T) f_k \otimes e_k^*$. Then, $\operatorname{rank} T_{n-1} = n-1$ and the equality $\|T - T_{n-1}\| = s_n(T)$ holds. $\quad\square$

Proposition 8.2. *For $S, T \in \mathbf{K}(\mathcal{H})$, $V \in \mathbf{B}(\mathcal{H})$, and $m, n \in \mathbb{N}$, the following hold:*

(1) $s_n(VT) \leq \|V\| s_n(T)$ *and* $s_n(TV) \leq \|V\| s_n(T)$.
(2) $s_{m+n-1}(S + T) \leq s_m(S) + s_n(T)$.

Proof. (1) We take $F \in \mathbf{F}(\mathcal{H})$ satisfying $\operatorname{rank} F \leq n-1$, and $s_n(T) = \|T - F\|$. Then, since $\operatorname{rank} VF \leq n-1$, we get

$$s_n(VT) \leq \|VT - VF\| \leq \|V\| \|T - F\| = \|V\| s_n(T).$$

The second inequality holds in the same way.

(2) We take $F' \in \mathbf{F}(\mathcal{H})$ satisfying $\operatorname{rank} F' \leq m-1$, and $s_m(S) = \|S - F'\|$. Then, since $\operatorname{rank}(F' + F) \leq m+n-2$, we have

$$s_{m+n-1}(S + T) \leq \|S + T - (F' + F)\| \leq \|S - F'\| + \|T - F\|$$

$$= s_m(S) + s_n(T). \qquad\square$$

Problem 8.1. Show the following *mini-max principle*; here, $\mathcal{K}$ runs over $n-1$-dimensional subspaces of $\mathcal{H}$:

(1) For $T \in \mathbf{K}(\mathcal{H})$,

$$s_n(T) = \min_{\dim \mathcal{K} = n-1} \max_{x \in \mathcal{K}^\perp, \|x\|=1} \|Tx\|.$$

(2) For $A \in \mathbf{K}(\mathcal{H})_+$,

$$s_n(A) = \min_{\dim \mathcal{K} = n-1} \max_{x \in \mathcal{K}^\perp, \|x\|=1} \langle Ax, x \rangle.$$

8.2 The Trace Class and the Hilbert–Schmidt Class

In this section, we assume that $\mathcal{H}$ is a countable infinite-dimensional Hilbert space for simplicity, though this assumption is not essential.

Definition 8.4. For $1 \leq p < \infty$, we let

$$\mathbf{S}_p(\mathcal{H}) = \left\{ T \in \mathbf{K}(\mathcal{H}); \sum_{n=1}^{\infty} s_n(T)^p < \infty \right\},$$

and call it the *Schatten p-class*. For $T \in \mathbf{S}_p(\mathcal{H})$, we define $\|T\|_p = \left(\sum_{n=1}^{\infty} s_n(T)^p\right)^{1/p}$ and call it the Schatten p-norm.

We call $\mathbf{S}_1(\mathcal{H})$ the *trace class* and write $\|\cdot\|_{\mathrm{Tr}} = \|\cdot\|_1$.

We call $\mathbf{S}_2(\mathcal{H})$ the *Hilbert–Schmidt class* and write $\|\cdot\|_{\mathrm{HS}} = \|\cdot\|_2$.

Lemma 8.2. *The Schatten class* $\mathbf{S}_p(\mathcal{H})$ *is a two-sided ideal of* $\mathbf{B}(\mathcal{H})$ *and* $\mathbf{K}(\mathcal{H})$, *and the inequalities*

$$\|VT\|_p \leq \|V\|\|T\|_p, \quad \|TV\|_p \leq \|T\|_p\|V\|,$$

hold for all $T \in \mathbf{S}_p(\mathcal{H})$ *and* $V \in \mathbf{B}(\mathcal{H})$. *Also, the following holds:* $T \in \mathbf{S}_p(\mathcal{H}) \iff T^* \in \mathbf{S}_p(\mathcal{H})$.

Proof. Since the singular numbers are monotone decreasing, the first statement follows from Proposition 8.2. The second statement follows from $s_n(T) = s_n(T^*)$. $\qquad\square$

It is impossible to show that the Schatten norm $\|\cdot\|_p$ satisfies the triangle inequality solely from Proposition 8.2 (see Gohberg and Krein (1969) and Simon (2005) for the proof). For applications, the cases of $p = 1, 2$ are particularly important, and we show this fact only for these cases. In both cases, its relationship to the trace of $\mathbf{B}(\mathcal{H})$ is important.

First, we introduce the trace only for positive operators.

Lemma 8.3. *Let* $A \in \mathbf{B}(\mathcal{H})_+$, *and let* $T \in \mathbf{B}(\mathcal{H})$.

(1) *We choose a CONS* $\{e_n\}_{n=1}^{\infty}$ *for* $\mathcal{H}$, *and we define the trace of* A *by* $\mathrm{Tr}\, A = \sum_{n=1}^{\infty} \langle Ae_n, e_n \rangle \in [0, \infty]$. *Then,* $\mathrm{Tr}\, A$ *does not depend on the choice of* $\{e_n\}_{n=1}^{\infty}$.

(2) *We have* $\mathrm{Tr}(T^*T) = \mathrm{Tr}(TT^*) \geq \|T\|^2$.

Proof. If $\{f_n\}_{n=1}^\infty$ is also a CONS for $\mathcal{H}$, the Parseval equality shows

$$\sum_{n=1}^\infty \langle T^*Te_n, e_n \rangle = \sum_{n=1}^\infty \|Te_n\|^2 = \sum_{n=1}^\infty \sum_{m=1}^\infty |\langle Te_n, f_m \rangle|^2$$

$$= \sum_{m=1}^\infty \sum_{n=1}^\infty |\langle e_n, T^*f_m \rangle|^2 = \sum_{m=1}^\infty \|T^*f_m\|^2$$

$$= \sum_{m=1}^\infty \langle TT^*f_m, f_m \rangle.$$

Thus, letting $T = |A|^{\frac{1}{2}}$, we get (1), and letting $e_n = f_n$, we get the first part of (2).

For every $x \in \mathcal{H}$ of norm 1, there exists a CONS $\{e_n\}_{n=1}^\infty$ for $\mathcal{H}$, with $e_1 = x$, and $\mathrm{Tr}(T^*T) \geq \langle T^*Tx, x \rangle = \|Tx\|^2$, which shows $\mathrm{Tr}(T^*T) \geq \|T\|^2$. $\qquad\square$

Lemma 8.4. $\mathrm{Tr}(T^*T) < \infty \implies T \in \mathbf{K}(\mathcal{H})$.

Proof. We choose a CONS $\{e_n\}_{n=1}^\infty$ for $\mathcal{H}$, and let E_n be the projection onto $\mathrm{span}\{e_k\}_{k=1}^n$. Then, since

$$\|T(I - E_n)\|^2 \leq \mathrm{Tr}((I - E_n)T^*T(I - E_n)) = \sum_{k=n+1}^\infty \langle T^*Te_k, e_k \rangle,$$

we get $\lim_{n\to\infty} \|T - TE_n\| = 0$. Thus, $TE_n \in \mathbf{F}(\mathcal{H})$ implies $T \in \mathbf{K}(\mathcal{H})$. $\qquad\square$

Theorem 8.3. *For $T \in \mathbf{B}(\mathcal{H})$, the following conditions are equivalent:*

(1) $\mathrm{Tr}(T^*T) < \infty$.
(2) $T \in \mathbf{S}_2(\mathcal{H})$.

When T satisfies the above equivalent conditions, we have $\mathrm{Tr}(T^*T) = \|T\|_{\mathrm{HS}}^2$.

Proof. (1) $\implies$ (2). Since $\mathrm{Tr}(T^*T) < \infty$, we have $T \in \mathbf{K}(\mathcal{H})$. Let $T = \sum_{n=1}^N s_n(T)f_n \otimes e_n^*$ be the Schmidt expansion of T. We take a CONS $\{e'_i\}_{i \in I}$ for $\ker T$. Since $\{e_n\}_{n=1}^N \cup \{e'_i\}_{i \in I}$ is a CONS for $\mathcal{H}$, we get

$$\mathrm{Tr}(T^*T) = \sum_{n=1}^N \|Te_n\|^2 + \sum_{i \in I} \|Te'_i\|^2 = \sum_{n=1}^N s_n(T)^2,$$

and $T \in \mathbf{S}_2(\mathcal{H})$. (2) $\Longrightarrow$ (1) also follows from the same computation.

$\square$

Remark 8.2. From the proof of Lemma 8.3 and the above theorem, we see that when $\{e_n\}_{n=1}^\infty$ is a CONS for $\mathcal{H}$,

$$\|T\|_{\mathrm{HS}} = \left(\sum_{m,n=1}^\infty |\langle Te_n, e_m \rangle|^2 \right)^{1/2}.$$

This is nothing but the ℓ^2-norm of the matrix elements of T.

Next, let us extend the domain of the trace.

Theorem 8.4. *For $S, T \in \mathbf{S}_2(\mathcal{H})$ and a CONS $\{e_n\}_{n=1}^\infty$ for $\mathcal{H}$,*

$$\sum_{n=1}^\infty \langle STe_n, e_n \rangle$$

absolutely converges, and its value does not depend on the choice of $\{e_n\}_{n=1}^\infty$. We denote it by $\mathrm{Tr}(ST)$ and call it the trace of ST. The equality $\mathrm{Tr}(ST) = \mathrm{Tr}(TS)$ holds.

Proof. Note that the equality

$$ST = \frac{1}{4} \sum_{k=0}^3 i^k (S + i^k T^*)(S + i^k T^*)^*$$

holds, as in the case of the polarization identity. Let $V_k = S + i^k T^*$. Then, $V_k \in \mathbf{S}_2(\mathcal{H})$. Since $|\langle STe_n, e_n \rangle| \leq 4^{-1} \sum_{k=0}^3 \langle V_k V_k^* e_n, e_n \rangle$, we have

$$\sum_{n=1}^\infty |\langle STe_n, e_n \rangle| \leq \frac{1}{4} \sum_{k=0}^3 \mathrm{Tr}(V_k V_k^*) < \infty.$$

Using $TS = 4^{-1} \sum_{k=0}^{3} i^k V_k^* V_k$, we conclude that

$$\operatorname{Tr}(ST) = \frac{1}{4} \sum_{k=0}^{3} i^k \operatorname{Tr}(V_k V_k^*) = \frac{1}{4} \sum_{k=0}^{3} i^k \operatorname{Tr}(V_k^* V_k) = \operatorname{Tr}(TS).$$

$\square$

Lemma 8.5. *For $T \in \mathbf{K}(\mathcal{H})$, the following conditions are equivalent:*

(1) *There exist $S_1, S_2 \in \mathbf{S}_2(\mathcal{H})$ satisfying $T = S_1 S_2$.*
(2) $T \in \mathbf{S}_1(\mathcal{H})$.

If T satisfies the above equivalent conditions, we have $|\operatorname{Tr} T| \leq \|T\|_{\operatorname{Tr}}$.

Proof. Let $T = \sum_{n=1}^{N} s_n(T) f_n \otimes e_n^*$ be the Schmidt expansion. Taking a CONS $\{e_i'\}_{i \in I}$ for $\ker T$, we get a CONS $\{e_n\}_{n=1}^{N} \cup \{e_i'\}_{i \in I}$ for $\mathcal{H}$. We use this CONS to compute the trace as follows.

(1) $\Longrightarrow$ (2). Let $T = W|T|$ be the polar decomposition of T. Then, we have $|T| = W^* S_1 S_2$ and $W^* S_1, S_2 \in \mathbf{S}_2(\mathcal{H})$, which shows $\operatorname{Tr}|T| < \infty$. Since $|T| = \sum_{n=1}^{N} s_n(T) e_n \otimes e_n^*$, we have $\operatorname{Tr}|T| = \sum_{n=1}^{N} s_n(T) = \|T\|_{\operatorname{Tr}}$, and $T \in \mathbf{S}_1(\mathcal{H})$.

(2) $\Longrightarrow$ (1). Let

$$S_1 = \sum_{n=1}^{N} s_n(T)^{\frac{1}{2}} f_n \otimes e_n^*, \quad S_2 = \sum_{n=1}^{N} s_n(T)^{\frac{1}{2}} e_n \otimes e_n^*.$$

Then, $S_1, S_2 \in \mathbf{S}_2(\mathcal{H})$ and $T = S_1 S_2$.

If T satisfies these equivalent conditions, we have $\operatorname{Tr} T = \sum_{n=1}^{N} s_n(T) \langle f_n, e_n \rangle$, and $|\operatorname{Tr} T| \leq \|T\|_{\operatorname{Tr}}$. $\square$

Theorem 8.5. *Let $T \in \mathbf{S}_1(\mathcal{H})$.*

(1) For every $V \in \mathbf{B}(\mathcal{H})$, the equality $\operatorname{Tr}(VT) = \operatorname{Tr}(TV)$ holds.
(2) The following holds:

$$\|T\|_{\operatorname{Tr}} = \max_{V \in \mathbf{B}(\mathcal{H}), \, \|V\| \leq 1} |\operatorname{Tr}(VT)| = \sup_{V \in \mathbf{K}(\mathcal{H}), \, \|V\| \leq 1} |\operatorname{Tr}(VT)|.$$

Proof. (1) Decomposing T as $T = S_1 S_2$, $S_1, S_2 \in \mathbf{S}_2(\mathcal{H})$, we get

$$\mathrm{Tr}(VT) = \mathrm{Tr}((VS_1)S_2) = \mathrm{Tr}(S_2(VS_1)) = \mathrm{Tr}((S_2V)S_1)$$

$$= \mathrm{Tr}(S_1(S_2V)) = \mathrm{Tr}(TV).$$

(2) Since $\|VT\|_{\mathrm{Tr}} \leq \|V\| \|T\|_{\mathrm{Tr}}$ holds by Lemma 8.2, if $\|V\| \leq 1$, then $|\mathrm{Tr}(VT)| \leq \|VT\|_{\mathrm{Tr}} \leq \|T\|_{\mathrm{Tr}}$. Let $T = W|T|$ be the polar decomposition. Then, $\|W\| \leq 1$ and $\mathrm{Tr}(W^*T) = \mathrm{Tr}(|T|) = \|T\|_{\mathrm{Tr}}$, which shows the first equality.

Let $T = \sum_{n=1}^{N} s_n(T) f_n \otimes e_n^*$ and $N \in \mathbb{N}_0 \cup \{\infty\}$, be the Schmidt expansion. If $N < \infty$, we have $W \in \mathbf{F}(\mathcal{H})$, and the second equality holds. If $N = \infty$, we let E_n be the projection onto $\mathrm{span}\{e_k\}_{k=1}^{n}$ and set $V_n = E_n W^*$. Then, $V_n \in \mathbf{F}(\mathcal{H})$ and $\|V_n\| = 1$. From

$$\mathrm{Tr}(V_n T) = \mathrm{Tr}(E_n|T|) = \sum_{k=1}^{n} s_k(T) \to \|T\|_{\mathrm{Tr}} \quad (n \to \infty),$$

we get the second equality. $\qquad \square$

From the above theorem, we can see that if we define $\varphi_T(K) = \mathrm{Tr}(KT)$ for $T \in \mathbf{S}_1(\mathcal{H})$ and $K \in \mathbf{K}(\mathcal{H})$, we have $\varphi_T \in \mathbf{K}(\mathcal{H})^*$ and $\|\varphi_T\| = \|T\|_{\mathrm{Tr}}$. Thus, we see that $(\mathbf{S}_1(\mathcal{H}), \|\cdot\|_{\mathrm{Tr}})$ is a normed space.

Proposition 8.3. *The map* $\Phi : \mathbf{S}_1(\mathcal{H}) \to \mathbf{K}(\mathcal{H})^*$, $T \mapsto \varphi_T$, *is an isometric isomorphism. In particular,* $(\mathbf{S}_1(\mathcal{H}), \|\cdot\|_{\mathrm{Tr}})$ *is a Banach space.*

Proof. It suffices to show that Φ is a surjection. Let $\varphi \in \mathbf{K}(\mathcal{H})^*$, and set $b_\varphi(x, y) = \varphi(x \otimes y^*)$ for $x, y \in \mathcal{H}$. Then, b_φ is a sesquilinear form. From

$$|b_\varphi(x, y)| \leq \|\varphi\| \|x \otimes y^*\| = \|\varphi\| \|x\| \|y\|,$$

the form b_φ is bounded, and there exists $T \in \mathbf{B}(\mathcal{H})$ satisfying $\varphi(x \otimes y^*) = \langle Tx, y \rangle$ for all $x, y \in \mathcal{H}$. Let $T = W|T|$ be the polar decomposition. Let $\{e_n\}_{n=1}^{\infty}$ be a CONS for $\mathcal{H}$. Then,

$$\varphi\left(\sum_{k=1}^{n}(e_k \otimes e_k^*)W^*\right) = \varphi\left(\sum_{k=1}^{n} e_k \otimes (We_k)^*\right) = \sum_{k=1}^{n}\langle Te_k, We_k\rangle$$

$$= \sum_{k=1}^{n}\langle |T|e_k, e_k\rangle$$

and

$$\sum_{k=1}^{n}\langle |T|e_k, e_k\rangle \leq \|\varphi\|\left\|\sum_{k=1}^{n}(e_k \otimes e_k^*)W^*\right\| \leq \|\varphi\|.$$

Since n is arbitrary, we get $T \in \mathbf{S}_1(\mathcal{H})$, and $\|T\|_{\mathrm{Tr}} \leq \|\varphi\|$ holds. Since

$$\varphi_T(x \otimes y^*) = \sum_{n=1}^{\infty}\langle (x \otimes y^*)Te_n, e_n\rangle = \sum_{n=1}^{\infty}\langle Te_n, y\rangle\langle x, e_n\rangle$$

$$= \sum_{n=1}^{\infty}\langle x, e_n\rangle\langle e_n, T^*y\rangle = \langle x, T^*y\rangle = \langle Tx, y\rangle$$

$$= \varphi(x \otimes y^*),$$

we have $\varphi|_{\mathbf{F}(\mathcal{H})} = \varphi_T|_{\mathbf{F}(\mathcal{H})}$. As $\mathbf{F}(\mathcal{H})$ is dense in $\mathbf{K}(\mathcal{H})$, we get $\varphi = \varphi_T$.
$\square$

Problem 8.2. For $V \in \mathbf{B}(\mathcal{H})$ and $T \in \mathbf{S}_1(\mathcal{H})$, we define $\psi_V(T) = \mathrm{Tr}(VT)$. Show that the map $\Psi : \mathbf{B}(\mathcal{H}) \to \mathbf{S}_1(\mathcal{H})^*$, $V \mapsto \psi_V$, is an isometric isomorphism.

Proposition 8.4. *For $S, T \in \mathbf{S}_2(\mathcal{H})$, we define $\langle S, T\rangle_{\mathrm{HS}} = \mathrm{Tr}(T^*S)$. Then, $(\mathbf{S}_2(\mathcal{H}), \langle \cdot, \cdot\rangle_{\mathrm{HS}})$ is a Hilbert space.*

Proof. We leave it to the reader to show that $\langle \cdot, \cdot\rangle_{\mathrm{HS}}$ satisfies the axiom of an inner product, and we only prove the completeness of $\mathbf{S}_2(\mathcal{H})$. Note that $\langle T, T\rangle_{\mathrm{HS}} = \|T\|_{\mathrm{HS}}^2$. Let $\{T_n\}_{n=1}^{\infty}$ be a Cauchy sequence in $(\mathbf{S}_2(\mathcal{H}), \|\cdot\|_{\mathrm{HS}})$. Since $\|T_m - T_n\|_{\mathrm{HS}} \geq \|T_m - T_n\|$, the sequence $\{T_n\}_{n=1}^{\infty}$ is a Cauchy sequence with respect to the operator norm as well. Thus, there exists $T \in \mathbf{B}(\mathcal{H})$ satisfying $\lim_{n\to\infty}\|T_n - T\| = 0$.

Let $\{e_k\}_{k=1}^{\infty}$ be a CONS for $\mathcal{H}$. As $\{T_n\}_{n=1}^{\infty}$ is a Cauchy sequence in $(\mathbf{S}_2(\mathcal{H}), \|\cdot\|_{\mathrm{HS}})$, the following holds:

$$\forall \varepsilon > 0, \ \exists N \in \mathbb{N}, \ \forall m > \forall n \geq N, \ \|T_m - T_n\|_{\mathrm{HS}} < \varepsilon.$$

Thus, for the above ε, m, n and an arbitrary $l \in \mathbb{N}$, we have

$$\sum_{k=1}^{l} \|(T_m - T_n)e_k\|^2 < \varepsilon^2.$$

Letting $m \to \infty$ here, we get

$$\sum_{k=1}^{l} \|(T - T_n)e_k\|^2 \leq \varepsilon^2,$$

and since l is arbitrary, we obtain $\|T - T_n\|_{\mathrm{HS}} \leq \varepsilon$. Therefore, we conclude $T = T - T_n + T_n \in \mathbf{S}_2(\mathcal{H})$ and $\lim_{n \to \infty} \|T - T_n\|_{\mathrm{HS}} = 0$. $\quad\square$

Corollary 8.4. *For any $S, T \in \mathbf{S}_2(\mathcal{H})$, the inequality $\|ST\|_{\mathrm{Tr}} \leq \|S\|_{\mathrm{HS}}\|T\|_{\mathrm{HS}}$ holds.*

Proof. Let $ST = W|ST|$ be the polar decomposition. Then, the Cauchy–Schwarz inequality shows

$$\|ST\|_{\mathrm{Tr}} = \mathrm{Tr}(W^*ST) = \langle T, S^*W \rangle_{\mathrm{HS}} \leq \|T\|_{\mathrm{HS}}\|S^*W\|_{\mathrm{HS}}$$

$$\leq \|T\|_{\mathrm{HS}}\|W\|\|S^*\|_{\mathrm{HS}} \leq \|S\|_{\mathrm{HS}}\|T\|_{\mathrm{HS}}. \quad\square$$

8.3   Hilbert–Schmidt Integral Operators

In this section, we assume that (Ω, μ) is a σ-finite measure space and $\mathcal{H} = L^2(\Omega, \mu)$ is separable. For measurable functions f and g on Ω, we write

$$(f \otimes g)(\xi, \eta) = f(\xi)g(\eta).$$

Let $k \in L^2(\Omega^2, \mu \otimes \mu)$. Since $\int_{\Omega} \int_{\Omega} |k(\xi, \eta)|^2 d\mu(\eta)d\mu(\xi) < \infty$, the Fubini theorem shows $\Omega_0 = \{\int_{\Omega} |k(\omega, \eta)|^2 d\mu(\eta) < \infty\}$ satisfies $\mu(\Omega \backslash \Omega_0) = 0$. Thus, for $\xi \in \Omega_0$ and $f \in \mathcal{H}$,

$$A_k f(\xi) = \int_{\Omega} k(\xi, \eta)f(\eta)d\mu(\eta)$$

makes sense.

Lemma 8.6. *We have $A_k \in \mathbf{B}(\mathcal{H})$, and $\|A_k\| \leq \|k\|_2$ holds.*

Proof. Since μ is σ-finite, there exists an increasing sequence of measurable sets of finite measure $\{E_n\}_{n=1}^{\infty}$ satisfying $\bigcup_{n=1}^{\infty} E_n = \Omega$. Since $k(\chi_{E_n} \otimes f)$ is integrable on Ω^2 for $f \in \mathcal{H}$, the Fubini theorem shows that

$$\Omega_0 \ni \xi \mapsto \int_\Omega k(\xi, \eta) \chi_{E_n}(\xi) f(\eta) d\mu(\eta) = \chi_{E_n}(\xi) A_k f(\xi)$$

is measurable. Since $\lim_{n \to \infty} \chi_{E_n}(\xi) A_k f(\xi) = A_k f(\xi)$, the function $A_k f(\xi)$ is measurable. By the Cauchy–Schwarz inequality, we get

$$|A_k f(\xi)|^2 = |\int_\Omega k(\xi, \eta) f(\eta) d\mu(\eta)|^2 \leq \int_\Omega |k(\xi, \eta)|^2 d\mu(\eta) \|f\|_2^2,$$

and $\int_\Omega |A_k f(\xi)|^2 d\mu(\xi) \leq \|k\|_2^2 \|f\|_2^2$ holds. $\qquad \square$

Definition 8.5. We call A_k the *Hilbert–Schmidt integral operator*, and call k its *integral kernel*. We set $k^*(\xi, \eta) = \overline{k(\eta, \xi)}$, which is called the *adjoint kernel* of k. We have $A_k{}^* = A_{k^*}$.

Theorem 8.6. *The operator A_k belongs to $\mathbf{S}_2(\mathcal{H})$, and $\|A_k\|_{\mathrm{HS}} = \|k\|_2$ holds.*

Proof. Let $\{e_n\}_{n=1}^{\infty}$ be a CONS for $\mathcal{H}$. Note that $\{\overline{e_n}\}_{n=1}^{\infty}$ is also a CONS. By the Parseval equality, for any $\xi \in \Omega_0$, we have

$$\int_\Omega |k(\xi, \eta)|^2 d\mu(\eta) = \sum_{n=1}^{\infty} |\langle k(\xi, \cdot), \overline{e_n} \rangle|^2 = \sum_{n=1}^{\infty} \left| \int_\Omega k(\xi, \eta) e_n(\eta) d\mu(\eta) \right|^2$$

$$= \sum_{n=1}^{\infty} |A_k e_n(\xi)|^2.$$

Thus, by the Fubini theorem,

$$\|k\|_2^2 = \int_\Omega \int_\Omega |k(\xi, \eta)|^2 d\mu(\eta) d\mu(\xi) = \sum_{n=1}^{\infty} \int_\Omega |A_k e_n(\xi)|^2 d\mu(\xi)$$

$$= \sum_{n=1}^{\infty} \|A_k e_n\|_2^2 = \|A_k\|_{\mathrm{HS}}^2.$$

$\qquad \square$

From the above theorem, we have $A_k \in \mathbf{S}_2(\mathcal{H})$, and A_k has the Schmidt expansion.

Theorem 8.7. *Let* $A_k = \sum_{n=1}^{N} s_n(A_k) f_n \otimes e_n^*$ *be the Schmidt expansion. Then,*

$$k = \sum_{n=1}^{N} s_n(A_k) f_n \otimes \overline{e_n}$$

holds in $L^2(\Omega^2, \mu \otimes \mu)$.

Proof.   Note that $\{f_n \otimes \overline{e_n}\}_{n=1}^{N}$ is an ONS in $L^2(\Omega^2, \mu \otimes \mu)$. From the previous theorem, we have $\sum_{n=1}^{N} s_n(A_k)^2 = \|A_k\|_{\mathrm{HS}}^2 = \|k\|_2^2 < \infty$, and $k' = \sum_{n=1}^{N} s_n(A_k) f_n \otimes \overline{e_n}$ converges in $L^2(\Omega^2, \mu \otimes \mu)$. Since

$$\|k - k'\|_2^2 = \|k\|_2^2 - 2\operatorname{Re}\langle k, k' \rangle + \|k'\|_2^2$$

$$= 2\sum_{n=1}^{N} s_n(A_k)^2 - 2\operatorname{Re}\sum_{n=1}^{N} s_n(A_k)\langle k, f_n \otimes \overline{e_n} \rangle,$$

and

$$\langle k, f_n \otimes \overline{e_n} \rangle = \int_\Omega \int_\Omega k(\xi, \eta)\overline{f_n(\xi)}e_n(\eta)d\mu(\eta)d\mu(\xi)$$

$$= \langle A_k e_n, f_n \rangle = s_n(A_k),$$

we get $k = k' \in L^2(\Omega^2, \mu \otimes \mu)$.   □

Remark 8.3. From the above proof, we see that every $T \in \mathbf{S}_2(L^2(\Omega, \mu))$ is a Hilbert–Schmidt integral operator. Indeed, we can construct an integral kernel k from the Schmidt expansion of T and show $T = A_k$.

Example 8.1. Let $\Omega = [0, 1]$, let μ be the Lebesgue measure, and let $k(s, t) = \min\{s, t\}$. Then, $k^* = k$, and $A_k \in \mathbf{S}_2(L^2[0, 1])$ is self-adjoint. Assume that $\lambda \in \mathbb{R}\backslash\{0\}$ and $f \in L^2[0, 1]\backslash\{0\}$ satisfy $A_k f = \lambda f$. Then, we have

$$\lambda f(s) = \int_0^s tf(t)dt + s\int_s^1 f(t)dt.$$

Since the right-hand side is continuous, so is the left-hand side, and f is continuous. Since the right-hand side is in $C^1[0, 1]$, so is the left-hand side, and f is in $C^1[0,1]$. Repeating this argument, we see that f is in $C^\infty[0,1]$. Differentiating both sides, we get $\lambda f'(s) = \int_s^1 f(t)dt$, and the above integral equation is equivalent to this condition, together with the boundary condition $f(0) = 0$. By differentiating both sides again, we see that the original integral equation is equivalent to

$$\lambda f''(s) = -f(s), \quad f(0) = f'(1) = 0.$$

Solving this, we can get the Schmidt expansion $A_k = \sum_{n=1}^\infty s_n(A_k)e_n \otimes e_n^*$, with $s_n(A_k) = 4\pi^{-2}(2n-1)^{-2}$ and $e_n(t) = \sqrt{2}\sin(n-1/2)\pi t$. In particular, A_k is a positive operator and $\|A_k\| = 4\pi^{-2}$.

Problem 8.3. We define the *Volterra operator* $V \in \mathbf{B}(L^2[0,1])$ by $Vf(s) = \int_0^s f(t)dt$. It is a Hilbert–Schmidt integral operator with the integral kernel

$$k(s,t) = \begin{cases} 1, & s \geq t, \\ 0, & s < t. \end{cases}$$

Describe the Schmidt expansion of V concretely. Also, find $\|V\|$.

8.4   Mercer's Theorem[#]

If the equality in Theorem 8.7 holds on the diagonal set of Ω^2 and we are allowed to perform termwise integration, and moreover, if $A_k \in \mathbf{S}_1(\mathcal{H})$, we would get

$$\int_\Omega k(\xi,\xi)d\mu(\xi) = \sum_{n=1}^N s_n(A_k)\langle f_n, e_n \rangle = \operatorname{Tr} A_k.$$

Of course, we cannot justify this argument for general k; even if Ω is a topological space and k is continuous, there is no guarantee that A_k belongs to the trace class (see Exercise 8.4). Mercer's theorem justifies this argument when k is a positive definite kernel. As it gives a concrete way to compute the trace, it is very useful.

Throughout this section, we assume that Ω is a compact metric space and μ is a finite Borel measure on Ω. Then, μ is regular and

$C(\Omega)$ is dense in $\mathcal{H} = L^2(\Omega, \mu)$, which is separable (see Section A.5). Furthermore, we assume that $\operatorname{supp}\mu = \Omega$ and $k \in C(\Omega^2)$. In this case, we have $\mathcal{R}(A_k), \mathcal{R}(A_{k^*}) \subset C(\Omega)$. Let $A_k = \sum_{n=1}^{N} s_n(A_k) f_n \otimes e_n^*$ be the Schmidt expansion. Since $A_k e_n = s_n(A_k) f_n$ and $A_{k^*} f_n = s_n(A_k) e_n$, we have $e_n, f_n \in C(\Omega)$. In what follows, we simply denote $s_n(A_k)$ by s_n if there is no possibility of confusion.

Theorem 8.8 (Schmidt theorem). *Let* $k \in C(\Omega^2)$, *and let* $A_k = \sum_{n=1}^{N} s_n(A_k) f_n \otimes e_n^*$ *be the Schmidt expansion. Then, for every* $g \in L^2(\Omega, \mu)$,

$$A_k g = \sum_{n=1}^{N} s_n(A_k) \langle g, e_n \rangle f_n$$

converges absolutely and uniformly.

Proof. We prove the theorem when $N = \infty$. It suffices to show the following: $\forall \varepsilon > 0$, $\exists L \in \mathbb{N}$, $\forall m > \forall l \geq L$, $\forall \xi \in \Omega$, $\sum_{n=l}^{m} |\langle g, e_n \rangle s_n f_n(\xi)| < \varepsilon$.

Since $\{\overline{e_n}\}_{n=1}^{\infty}$ is an ONS of $\mathcal{H}$, the Bessel inequality shows

$$\sum_{n=1}^{\infty} |\langle k(\xi, \cdot), \overline{e_n} \rangle|^2 \leq \int_{\Omega} |k(\xi, \eta)|^2 d\mu(\eta) \leq \|k\|_{\infty}^2 \mu(\Omega).$$

Here, we have

$$\langle k(\xi, \cdot), \overline{e_n} \rangle = \int_{\Omega} k(\xi, \eta) e_n(\eta) d\mu(\eta) = A_k e_n(\xi) = s_n f_n(\xi),$$

and $\sum_{n=1}^{\infty} s_n^2 |f_n(\xi)|^2 \leq \|k\|_{\infty}^2 \mu(\Omega)$ holds. For $1 \leq l < m$,

$$\sum_{n=l}^{m} |\langle g, e_n \rangle s_n f_n(\xi)| \leq \left(\sum_{n=l}^{m} |\langle g, e_n \rangle|^2 \right)^{\frac{1}{2}} \left(\sum_{n=l}^{m} s_n^2 |f_n(\xi)|^2 \right)^{\frac{1}{2}}$$

$$\leq \left(\sum_{n=l}^{m} |\langle g, e_n \rangle|^2 \right)^{\frac{1}{2}} \|k\|_{\infty} \sqrt{\mu(\Omega)}$$

holds. The Bessel inequality shows that $\sum_{n=1}^{\infty} |\langle g, e_n \rangle|^2$ converges, and the statement follows. $\qquad\square$

Definition 8.6. We say that a function $k \in C(\Omega^2)$ is a *positive definite kernel* if for all $n \in \mathbb{N}$, for all n distinct points $\omega_1, \omega_2, \ldots, \omega_n \in \Omega$, and for all $c_1, c_2, \ldots, c_n \in \mathbb{C}$, we have $\sum_{i,j=1}^{n} k(\omega_i, \omega_j) c_i \overline{c_j} \geq 0$.

When k is a positive definite kernel, we can see the following from the definition:

(1) The condition in the case of $n = 1$ shows $k(\omega, \omega) \geq 0$ for every $\omega \in \Omega$.
(2) Proposition 2.3 shows that $(k(\omega_i, \omega_j))_{i,j}$ is a Hermitian matrix and $k = k^*$ holds. In particular, A_k is self-adjoint.

Example 8.2. For $\Omega = [0,1]$, the function $k(s,t) = \min\{s,t\}$ is a positive definite kernel. This follows from $\min\{s,t\} = \langle \chi_{[0,s]}, \chi_{[0,t]} \rangle$.

Lemma 8.7. *For $k \in C(\Omega^2)$, the following conditions are equivalent:*

(1) *k is a positive definite kernel.*
(2) *A_k is a positive operator.*

Proof. (1) $\Longrightarrow$ (2). Since k is uniformly continuous on Ω^2, for every $\varepsilon > 0$, there exists $\delta > 0$ such that if $(\xi, \eta), (\xi', \eta') \in \Omega^2$ satisfy $d(\xi, \xi') < \delta$ and $d(\eta, \eta') < \delta$, we have $|k(\xi, \eta) - k(\xi', \eta')| < \varepsilon$. Since Ω is compact, there exist $\omega_1, \omega_2, \ldots, \omega_n \in \Omega$ satisfying $\Omega = \bigcup_{i=1}^{n} B(\omega_i, \delta)$. We choose a partition of unity $\{h_i\}_{i=1}^{n}$ subordinate to this open cover (see Folland, 1999). That is, the functions $h_i \in C(\Omega)$ satisfy $0 \leq h_i \leq 1$, $\sum_{i=1}^{n} h_i = 1$, and $\operatorname{supp} h_i \subset B(\omega_i, \delta)$. For $f \in \mathcal{H}$, let $c_i = \int_{\Omega} h_i(\omega) f(\omega) d\mu(\omega)$. Then, the following holds:

$$\left| \langle A_k f, f \rangle - \sum_{i,j=1}^{n} k(\omega_i, \omega_j) c_i \overline{c_j} \right|$$

$$= \left| \sum_{i,j=1}^{n} \int_{\Omega^2} (k(\xi, \eta) - k(\omega_i, \omega_j)) h_i(\xi) f(\xi) \overline{h_j(\eta) f(\eta)} d\mu(\xi) d\mu(\eta) \right|$$

$$\leq \varepsilon \sum_{i,j=1}^{n} \int_{\Omega^2} h_i(\xi) h_j(\eta) |f(\xi) f(\eta)| d\mu(\xi) d\mu(\eta)$$

$$= \varepsilon \|f\|_1^2.$$

Thus,

$$\langle A_k f, f \rangle \geq \sum_{i,j=1}^{n} k(\omega_i, \omega_j) c_i \overline{c_j} - \varepsilon \mu(\Omega) \|f\|_2^2 \geq -\varepsilon \mu(\Omega) \|f\|_2^2$$

holds. As $\varepsilon > 0$ is arbitrary, we get $\langle A_k f, f \rangle \geq 0$.

(2) $\implies$ (1). Let $\omega_1, \omega_2, \ldots, \omega_n$ be n distinct points in Ω, and let $c_1, c_2, \ldots, c_n \in \mathbb{C}$. For an arbitrary $\varepsilon > 0$, we choose $f_1, f_2, \ldots, f_n \in C(\Omega)$ satisfying $0 \leq f_i$, $\operatorname{supp} f_i \subset B(\omega_i, \varepsilon)$, and $\int_\Omega f_i d\mu = 1$. Let $f = \sum_{i=1}^{n} c_i f_i$. Then,

$$0 \leq \langle A_k f, f \rangle = \sum_{i,j=1}^{n} c_i \overline{c_j} \int_{\Omega^2} k(\xi, \eta) f_i(\xi) f_j(\eta) d\mu(\xi) d\mu(\eta).$$

When $\varepsilon \downarrow 0$, the right-hand side tends to $\sum_{i,j=1}^{n} k(\omega_i, \omega_j) c_i \overline{c_j}$, and k is a positive definite kernel. $\qquad\square$

Theorem 8.9 (Mercer's theorem). *Let $k \in C(\Omega^2)$ be a positive definite kernel, and let $A_k = \sum_{n=1}^{N} s_n(A_k) e_n \otimes e_n^*$ be the Schmidt expansion. Then,*

$$k(\omega, \omega) = \sum_{n=1}^{N} s_n(A_k) |e_n(\omega)|^2$$

uniformly converges on Ω. The operator A_k belongs to $\mathbf{S}_1(\mathcal{H})_+$, and $\operatorname{Tr} A_k = \int_\Omega k(\omega, \omega) d\mu(\omega)$ holds.

Proof. We show the statement for $N = \infty$. For $m \in \mathbb{N}$, let $k_m(\xi, \eta) = \sum_{n=1}^{m} s_n e_n(\xi) \overline{e_n(\eta)}$. Then,

$$A_{k-k_m} = A_k - A_{k_m} = \sum_{n=m+1}^{\infty} s_n e_n \otimes e_n^*$$

is a positive operator, and $k - k_m$ is a positive definite kernel. Thus, for each $\xi \in \Omega$, we have $k(\xi, \xi) - k_m(\xi, \xi) \geq 0$, and $\sum_{n=1}^{m} s_n |e_n(\xi)|^2 \leq k(\xi, \xi)$ holds. As m is arbitrary, we get $\sum_{n=1}^{\infty} s_n |e_n(\xi)|^2 \leq k(\xi, \xi)$. This shows the following: $\forall \varepsilon > 0$, $\forall \xi \in \Omega$, $\exists L \in \mathbb{N}$, $\forall m > \forall l \geq L$,

$\sum_{n=l}^{m} s_n |e_n(\xi)|^2 < \varepsilon$. From this estimate, we see that for a fixed $\xi \in \Omega$,

$$k'(\xi, \eta) = \sum_{n=1}^{\infty} s_n e_n(\xi) \overline{e_n(\eta)}$$

uniformly converges as functions in η on Ω. Indeed, this is because

$$\left| \sum_{n=l}^{m} s_n e_n(\xi) \overline{e_n(\eta)} \right| \leq \left(\sum_{n=l}^{m} s_n |e_n(\xi)|^2 \right)^{\frac{1}{2}} \left(\sum_{n=l}^{m} s_n |e_n(\eta)|^2 \right)^{\frac{1}{2}}$$

$$\leq (\varepsilon \|k\|_\infty)^{\frac{1}{2}}.$$

From the above argument, the function $\Omega \ni \eta \mapsto k'(\xi, \eta)$ is continuous, and for every $g \in C(\Omega)$, we have

$$\int_\Omega k'(\xi, \eta) g(\eta) d\mu(\eta) = \sum_{n=1}^{\infty} s_n e_n(\xi) \int_\Omega g(\eta) \overline{e_n(\eta)} d\mu(\eta)$$

$$= \sum_{n=1}^{\infty} \langle g, e_n \rangle s_n e_n(\xi).$$

Since the right-hand side uniformly converges as functions in ξ on Ω to $A_k g(\xi) = \int_\Omega k(\xi, \eta) g(\eta) d\mu(\eta)$, the equality $k(\xi, \eta) = k'(\xi, \eta)$ holds for all $(\xi, \eta) \in \Omega^2$. In particular, $k(\omega, \omega) = \sum_{n=1}^{\infty} s_n |e_n(\omega)|^2$ converges pointwise. Since the left-hand side is continuous in ω and $|e_n(\omega)|^2$ is also continuous, the Dini theorem implies that the convergence is uniform on Ω. Integrating both sides, we get

$$\int_\Omega k(\omega, \omega) d\mu(\omega) = \sum_{n=1}^{\infty} s_n \int_\Omega |e_n(\omega)|^2 d\mu(\omega) = \sum_{n=1}^{\infty} s_n = \operatorname{Tr} A_k.$$

$\square$

Problem 8.4. By computing the trace of the operator in Example 8.1 in two ways, show the equality

$$\sum_{n=1}^{\infty} \frac{1}{(2n-1)^2} = \frac{\pi^2}{8}.$$

Exercises

Exercise 8.1
Show that a compact normal operator T is a diagonal operator in the following steps:

(1) Let $T = W|T|$ be the polar decomposition. Show that W commutes with $|T|$.
(2) Let $|T| = \sum_{n=1}^{N} \lambda_n P_n$, $N \in \mathbb{N}_0 \cup \{\infty\}$, be the spectral decomposition. Show that W commutes with P_n.
(3) Show that T is a diagonal operator.

Exercise 8.2
Let $\{T_n\}_{n=1}^{\infty}$ be a sequence in $\mathbf{B}(\mathcal{H})$, let $T \in \mathbf{B}(\mathcal{H})$, and let $K \in \mathbf{K}(\mathcal{H})$. Show the following:

(1) If $\{T_n\}_{n=1}^{\infty}$ weakly converges to T, the sequence $\{KT_n\}_{n=1}^{\infty}$ strongly converges to KT.
(2) If $\{T_n\}_{n=1}^{\infty}$ strongly converges to T, the sequence $\{T_n K\}_{n=1}^{\infty}$ converges to TK in the operator norm.

Exercise 8.3
For the Volterra operator V and $0 < \theta < \pi$, let $B_\theta = e^{i\theta} V + e^{-i\theta} V^*$. Find $\max_{\|f\|_2=1} \langle B_\theta f, f \rangle$.

Exercise 8.4
In the following exercises, we use the notation in Example 2.8 for the Fourier series. For $f \in L^2(\mathbb{T})$, we define an integral kernel $k \in L^2(\mathbb{T}^2)$ by $k(s,t) = f(s-t)$.

(1) Show that the integral operator A_k is a diagonal operator.
(2) Show that there exists $f \in C(\mathbb{T})$ such that $A_k \notin \mathbf{S}_1(L^2(\mathbb{T}))$.

Exercise 8.5
For $s \geq 0$, let $W^{s,2}(\mathbb{T}) = \{f \in L^2(\mathbb{T}); \sum_{n\in\mathbb{Z}}(1+n^2)^s|\hat{f}(n)|^2 < \infty\}$, which is called the *Sobolev space*. With the inner product

$$\langle f, g \rangle_s = \sum_{n\in\mathbb{Z}} (1+n^2)^s \, \hat{f}(n)\overline{\hat{g}(n)},$$

$W^{s,2}(\mathbb{T})$ is a Hilbert space. We write $\|f\|_{s,2} = \sqrt{\langle f, f \rangle_s}$.

(1) For $0 \le s_1 < s_2$, show that the embedding map $W^{s_2,2}(\mathbb{T}) \hookrightarrow W^{s_1,2}(\mathbb{T})$ is a compact operator from $(W^{s_2,2}, \langle \cdot, \cdot \rangle_{s_2})$ into $(W^{s_1,2}, \langle \cdot, \cdot \rangle_{s_1})$.

(2) For $m \in \mathbb{N}_0$ and $s > m + \frac{1}{2}$, show $W^{s,2}(\mathbb{T}) \subset C^m(\mathbb{T}) \subset W^{m,2}(\mathbb{T})$.

Exercise 8.6

In this exercise, we use the notation in Exercise 2.1. For $f, g \in L^\infty(\mathbb{T})$, show the following:

(1) $[M_f, P_+] \in \mathbf{S}_2(L^2(\mathbb{T})) \iff f \in W^{1/2,2}(\mathbb{T})$.

(2) If $f, g \in W^{1/2,2}(\mathbb{T})$, we have $[T_f, T_g] \in \mathbf{S}_1(H^2(\mathbb{T}))$ and $\|[T_f, T_g]\|_{\mathrm{Tr}} \le 2\|f\|_{1/2,2} \|g\|_{1/2,2}$.

(3) For $f \in C^1(\mathbb{T})$, $g \in W^{\frac{1}{2},2}(\mathbb{T})$, the following holds:

$$\mathrm{Tr}([T_f, T_g]) = \frac{-1}{2\pi i} \int_0^{2\pi} f'(t)g(t)dt.$$

Exercise 8.7

Show, in the following steps, that if $T, S \in \mathbf{B}(\mathcal{H})$ satisfy $I - ST$, $I - TS \in \mathbf{S}_1(\mathcal{H})$, we have $\mathrm{Tr}([T, S]) = \mathrm{ind}(T)$:

(1) Show that there exists $S_0 \in \mathbf{B}(\mathcal{H})$ satisfying $TS_0 = P_{\mathcal{R}(T)}$ and $S_0 T = I - P_{\ker T}$.

(2) Let $\rho : \mathbf{B}(\mathcal{H}) \to \mathbf{B}(\mathcal{H})/\mathbf{S}_1(\mathcal{H})$ be the quotient map. Show $\rho(S) = \rho(S_0)$.

(3) Show $\mathrm{Tr}([T, S]) = \mathrm{Tr}([T, S_0]) = \mathrm{ind}(T)$.

Exercise 8.8

By computing

$$\mathrm{Tr}\left(\left[T_f, T_{\frac{1}{f}}\right]\right) = \frac{-1}{2\pi i} \int_0^{2\pi} \frac{f'(t)}{f(t)} dt$$

for $f \in C^1(\mathbb{T}) \cap C(\mathbb{T})^{-1}$, show the index formula for the Toeplitz operator $\mathrm{ind}(T_f) = -w(f)$ again (see Exercise 7.5).

Chapter 9

Spectral Decomposition of Bounded Self-Adjoint Operators

In this chapter, we prove the spectral decomposition theorem for bounded self-adjoint operators and unitary operators. Unlike the case of compact operators, its formulation requires integration. As an application of the spectral decomposition, we generalize the continuous functional calculus to the Borel functional calculus.

9.1 The Spectral Decomposition Theorem for Bounded Self-Adjoint Operators

We assume that $\mathcal{H}$ is a Hilbert space throughout this chapter.

9.1.1 *Monotone convergence of a sequence of operators*

First, we make preparations for the monotone convergence limit of a sequence of operators. Since $\mathbf{B}(\mathcal{H})_{\mathrm{sa}}$ is an ordered set, it makes sense to define an upper bound of a subset $\mathcal{S}$. When there exists a smallest element among the upper bounds of $\mathcal{S}$, we call it the supremum of $\mathcal{S}$ and denote it by $\sup \mathcal{S}$. We define $\inf S$ in a similar way.

Theorem 9.1. *Let $\{A_n\}_{n=1}^\infty$ be a monotone increasing sequence in $\mathbf{B}(\mathcal{H})_{\mathrm{sa}}$, and assume that there exists $B \in \mathbf{B}(\mathcal{H})_{\mathrm{sa}}$ satisfying $A_n \leq B$ for all $n \in \mathbb{N}$. Then, there exists $A = \sup\{A_n\}_{n=1}^\infty \in \mathbf{B}(\mathcal{H})_{\mathrm{sa}}$, and $\{A_n\}_{n=1}^\infty$ strongly converges to A.*

Proof. Replacing A_n and B with $A_n - A_1$ and $B - A_1$, respectively, we may assume that $0 \le A_n \le B$ to prove the theorem. For every $x \in \mathcal{H}$, the sequence $\{\langle A_n x, x \rangle\}_{n=1}^{\infty}$ is bounded and monotone increasing, and it converges. From the polarization identity, for all x, y, the sequence $\{\langle A_n x, y \rangle\}_{n=1}^{\infty}$ converges, and we denote its limit by $f(x, y)$. Then, f is a sesquilinear form. For $\|x\|, \|y\| \le 1$, we have

$$|\langle A_n x, y \rangle| = \frac{1}{4} \left| \sum_{k=0}^{3} i^k \langle A_n (x + i^k y), x + i^k y \rangle \right|$$

$$\le \frac{1}{4} \sum_{k=0}^{3} \langle A_n (x + i^k y), x + i^k y \rangle$$

$$= \langle A_n x, x \rangle + \langle A_n y, y \rangle \le \langle Bx, x \rangle + \langle By, y \rangle \le 2\|B\|,$$

and f is bounded. Thus, there exists $A \in \mathbf{B}(\mathcal{H})$ such that $f(x, y) = \langle Ax, y \rangle$ holds for all $x, y \in \mathcal{H}$. By construction, the sequence $\{A_n\}_{n=1}^{\infty}$ weakly converges to A, and $A_n \le A \le B$ holds for every $n \in \mathbb{N}$. Thus, we get $A = \sup\{A_n\}_{n=1}^{\infty}$.

For every $x \in \mathcal{H}$, we have

$$\|Ax - A_n x\|^2 = \langle (A - A_n)(A - A_n)^{\frac{1}{2}} x, (A - A_n)^{\frac{1}{2}} x \rangle$$

$$\le \|A - A_n\| \|(A - A_n)^{\frac{1}{2}} x\|^2 \le \|A\| \langle (A - A_n) x, x \rangle,$$

and $\{A_n\}_{n=1}^{\infty}$ strongly converges to A. $\qquad\square$

In a similar way, if $\{A_n\}_{n=1}^{\infty}$ is a monotone decreasing sequence in $\mathbf{B}(\mathcal{H})_{\mathrm{sa}}$, and there exists $B \in \mathbf{B}(\mathcal{H})_{\mathrm{sa}}$ satisfying $B \le A_n$ for all $n \in \mathbb{N}$, then $\{A_n\}_{n=1}^{\infty}$ strongly converges to $\inf\{A_n\}_{n=1}^{\infty}$.

Assume that $\{P_n\}_{n=1}^{\infty}$ is a monotone increasing sequence of projections. Since $0 \le P_n \le I$, the strong limit $P = \text{s-}\lim_{n \to \infty} P_n$ exists, and Proposition 4.2 shows that P is a projection. In fact, we can show that P is the projection onto $\overline{\bigcup_{n=1}^{\infty} \mathcal{R}(P_n)} = \mathcal{K}$. Since $P_n \le P$ for all n, we have $\mathcal{R}(P_n) \subset \mathcal{R}(P)$, and $\mathcal{K} \subset \mathcal{R}(P)$. On the other hand, since $P_n \le P_{\mathcal{K}}$ for every n, we have $P \le P_{\mathcal{K}}$, and Lemma 6.13 implies $\mathcal{R}(P) \subset \mathcal{K}$. In a similar way, if $\{P_n\}_{n=1}^{\infty}$ is a monotone decreasing sequence of projections, it strongly converges to the projection onto $\bigcap_{n=1}^{\infty} \mathcal{R}(P_n)$.

Problem 9.1. Let $A \in \mathbf{B}(\mathcal{H})_+$.

(1) Show that $\{A^{1/n}\}_{n=1}^{\infty}$ strongly converges to the projection onto $\ker A^{\perp}$.
(2) When $A \leq I$, show that $\{A^n\}_{n=1}^{\infty}$ strongly converges to the projection onto $\ker(I - A)$.

9.1.2   *Construction of a spectral family*

When the spectrum of $A \in \mathbf{B}(\mathcal{H})_{\mathrm{sa}}$ is a finite set, $\sigma(A) = \{\lambda_1, \lambda_2, \ldots, \lambda_n\}$, Proposition 6.3 shows that we have the spectral decomposition $A = \sum_{i=1}^{n} \lambda_i P_{\lambda_i}$. However, for general $A \in \mathbf{B}(\mathcal{H})_{\mathrm{sa}}$, we cannot expect the same type of decomposition, and we rephrase it as follows. Let $E_\lambda^A = \sum_{\lambda_i \leq \lambda} P_{\lambda_i}$. Then, $\{E_\lambda^A\}_{\lambda \in \mathbb{R}}$ is a monotone increasing family of projections, and

$$\langle Ax, y \rangle = \sum_{i=1}^{n} \lambda_i \langle P_{\lambda_i} x, y \rangle = \int_{\mathbb{R}} t d\langle E_t^A x, y \rangle$$

holds in the sense of the Riemann–Stieltjes integral. We show in the following that the spectral decomposition of this form holds for general $A \in \mathbf{B}(\mathcal{H})_{\mathrm{sa}}$. The main difficulty is to construct $E_\lambda^A = \chi_{(-\infty, \lambda]}(A)$; we cannot use the continuous functional calculus because, in general, the function $\chi_{(-\infty, \lambda]}(t)$ is not continuous on $\sigma(A)$. However, this function is upper semi-continuous, and it can be approximated by continuous functions from above in the sense of pointwise convergence.

Problem 9.2. Let F be a closed subset of a metric space (Ω, d). Show that there exists a monotone decreasing sequence of continuous real-valued functions on Ω converging to χ_F pointwise.

Definition 9.1. We say that a family $E = \{E_\lambda\}_{\lambda \in \mathbb{R}}$ of projections on $\mathcal{H}$ is a *spectral family* if the following conditions are satisfied:

(1) If $\lambda_1 < \lambda_2$, we have $E_{\lambda_1} \leq E_{\lambda_2}$.
(2) For every $\lambda \in \mathbb{R}$, we have s-$\lim_{t \to \lambda+0} E_t = E_\lambda$.
(3) s-$\lim_{\lambda \to -\infty} E_\lambda = 0$ and s-$\lim_{\lambda \to \infty} E_\lambda = I$.

Example 9.1. The above $E^A = \{E_\lambda^A\}_{\lambda \in \mathbb{R}}$ is a spectral family.

Example 9.2. Let $\mathcal{H} = L^2(\mathbb{R})$, and let E_λ be the multiplication operator of the function $\chi_{(-\infty,\lambda]}(t)$. Then, $\{E_\lambda\}_{\lambda \in \mathbb{R}}$ is a spectral family.

We fix $A \in \mathbf{B}(\mathcal{H})_{\mathrm{sa}}$ in the following arguments and set $m = \min \sigma(A)$, $M = \max \sigma(A)$. For a closed subset F of $\sigma(A)$, let C_F be the set of real-valued continuous functions f on $\sigma(A)$ satisfying $\chi_F(t) \le f(t)$ for all $t \in \sigma(A)$.

Lemma 9.1. *We choose a monotone decreasing sequence of functions $\{f_n\}_{n=1}^\infty$ in C_F converging to χ_F pointwise, and we set $E^A(F) = $ s-$\lim_{n\to\infty} f_n(A)$. Then, $E^A(F) = \inf\{f(A);\ f \in C_F\}$ holds. In particular, the definition of $E^A(F)$ does not depend on the choice of $\{f_n\}_{n=1}^\infty$, and $E^A(F)$ is a projection. If $T \in \mathbf{B}(\mathcal{H})$ commutes with A, it commutes with $E^A(F)$ as well.*

Proof. We first show the following claim: $\forall f \in C_F$, $\forall \varepsilon > 0$, $\exists N \in \mathbb{N}$, $\forall n \ge N$, $\forall t \in \sigma(A)$, $f_n(t) < f(t) + \varepsilon$. Indeed, let

$$F_n = \{t \in \sigma(A);\ f(t) + \varepsilon \le f_n(t)\}.$$

Then, $\{F_n\}_{n=1}^\infty$ is a monotone decreasing sequence of compact sets with $\bigcap_{n=1}^\infty F_n = \emptyset$. Thus, there exists N satisfying $F_N = \emptyset$, and the claim holds. The claim shows that $E^A(F) \le f(A) + \varepsilon I$, and since $\varepsilon > 0$ is arbitrary, we get $E^A(F) \le f(A)$. This shows that $E^A(F)$ is a lower bound of $\{f(A);\ f \in C_F\}$. Since $E^A(F) = \inf\{f_n(A)\}$, we get $E^A(F) = \inf\{f(A);\ f \in C_F\}$.

Since $\{f_n^2\}_{n=1}^\infty$ also satisfies the same condition as $\{f_n\}_{n=1}^\infty$, we have s-$\lim_{n\to\infty} f_n(A)^2 = E^A(F)$, and $E^A(F)^2 = E^A(F)$ holds. Thus, $E^A(F)$ is a projection. $\qquad\square$

If F_1 and F_2 are closed subsets of $\sigma(A)$ with $F_1 \subset F_2$, we have $C_{F_2} \subset C_{F_1}$, and $\inf\{f(A);\ f \in C_{F_1}\} \le \inf\{f(A);\ f \in C_{F_2}\}$. Thus, $E^A(F_1) \le E^A(F_2)$ holds.

Lemma 9.2. *Let $\{F_n\}_{n=1}^\infty$ be a decreasing sequence of closed subsets of $\sigma(A)$, with $F = \bigcap_{n=1}^\infty F_n$. Then, s-$\lim_{n\to\infty} E^A(F_n) = E^A(F)$ holds.*

Proof.   Since $\{E^A(F_n)\}_{n=1}^{\infty}$ is a decreasing sequence of projections, it strongly converges to a projection P. Since $E^A(F_n) \geq E^A(F)$ holds for all $n \in \mathbb{N}$, we have $P \geq E^A(F)$.

Let $f \in C_F$, and let $\varepsilon > 0$. Then, there exists $N \in \mathbb{N}$ with $f + \varepsilon \in C_{F_N}$. Indeed, since $\{t \in F_n; \, f(t) + \varepsilon \leq 1\}_{n=1}^{\infty}$ is a decreasing sequence of compact sets whose intersection is $\{t \in F; \, f(t) + \varepsilon \leq 1\} = \emptyset$, there exists $N \in \mathbb{N}$ satisfying $\{t \in F_N; \, f(t) + \varepsilon \leq 1\} = \emptyset$. This shows $P \leq E^A(F_N) \leq f(A) + \varepsilon I$. Since $\varepsilon > 0$ is arbitrary, we get $P \leq f(A)$. Since this is the case for every $f \in C_F$, we get $P \leq E^A(F)$. $\square$

In what follows, when F is a closed subset of $\mathbb{R}$, the symbol C_F means $C_{\sigma(A) \cap F}$ and $E^A(F)$ means $E^A(\sigma(A) \cap F)$ for simplicity.

Corollary 9.1. *For $A \in \mathbf{B}(\mathcal{H})_{\mathrm{sa}}$ and $\lambda \in \mathbb{R}$, let $E_{\lambda}^A = E^A((-\infty, \lambda])$. Then, $E^A = \{E_{\lambda}^A\}_{\lambda \in \mathbb{R}}$ is a spectral family. For $\lambda < m$, we have $E_{\lambda}^A = 0$, and for $M \leq \lambda$, we have $E_{\lambda}^A = I$.*

For $\lambda \in \mathbb{R}$ and $n \in \mathbb{N}$, we define $f_{\lambda, n} \in C_{(-\infty, \lambda]}$ by

$$f_{\lambda,n}(t) = \begin{cases} 1, & t \leq \lambda, \\ n\lambda + 1 - nt, & \lambda < t \leq \lambda + \dfrac{1}{n}, \\ 0, & \lambda + \dfrac{1}{n} < t. \end{cases}$$

Then, $\{f_{\lambda,n}\}_{n=1}^{\infty}$ is a monotone decreasing sequence of continuous functions converging to $\chi_{(-\infty,\lambda]}$ pointwise, and $E_{\lambda}^A =$ s-$\lim_{n \to \infty} f_{\lambda,n}(A)$.

For $\mu < \lambda$, let $E^A((\mu, \lambda]) = E_{\lambda}^A - E_{\mu}^A$. Then, $E^A((\mu, \lambda])$ is the projection onto $\mathcal{R}(E_{\lambda}^A) \cap \mathcal{R}(E_{\mu}^A)^{\perp}$.

Lemma 9.3. *For all $\mu < \lambda$ and $f \in C(\sigma(A))$, we have*

$$\|f(A)E^A((\mu, \lambda])\| \leq \sup_{t \in \sigma(A) \cap [\mu, \lambda]} |f(t)|.$$

Proof.   Let $g_n(t) = f_{\lambda,n}(t) - f_{\mu,n}(t)$. Then, we have $E^A((\mu, \lambda]) =$ s-$\lim_{n \to \infty} g_n(A)$. Note that we have $\mathrm{supp}\, g_n = [\mu, \lambda + \frac{1}{n}]$ and

$0 \leq g_n(t) \leq 1$. Thus, for every $x \in \mathcal{H}$,

$$\|f(A)E^A((\mu, \lambda])x\| = \lim_{n \to \infty} \|f(A)g_n(A)x\| \leq \lim_{n \to \infty} \|f(A)g_n(A)\|\|x\|,$$

and

$$\|f(A)E^A((\mu, \lambda])\| \leq \lim_{n \to \infty} \sup_{t \in \sigma(A) \cap [\mu, \lambda + \frac{1}{n}]} |f(t)| = \sup_{t \in \sigma(A) \cap [\mu, \lambda]} |f(t)|$$

holds. □

We fix $a < m$ and extend f to $[a, M]$ continuously. Let $\Delta : a = t_0 < t_1 < \cdots < t_n = M$ be a partition of the closed interval $[a, M]$, and we define the mesh of Δ by $h(\Delta) = \max_{1 \leq i \leq n}(t_i - t_{i-1})$. For $\xi_i \in [t_{i-1}, t_i]$, $i = 1, 2, \ldots, n$, we set

$$S(f, \Delta, \{\xi_i\}, E^A) = \sum_{i=1}^{n} f(\xi_i)E^A((t_{i-1}, t_i]).$$

Theorem 9.2 (Spectral decomposition). *For every $f \in C(\sigma(A))$, we have*

$$\lim_{h(\Delta) \to 0} \|S(f, \Delta, \{\xi_i\}, E^A) - f(A)\| = 0.$$

In particular, for any $x, y \in \mathcal{H}$,

$$\langle f(A)x, y \rangle = \int_{m-0}^{M} f(t)d\langle E_t^A x, y \rangle$$

holds as Riemann–Stieltjes integral (see Section A.6).

Proof. For $\delta > 0$, let

$$m(f, \delta) = \sup\{|f(s) - f(t)|;\ s, t \in [a, M],\ |s - t| \leq \delta\}.$$

Since f is uniformly continuous on $[a, M]$, we have $\lim_{\delta \to +0} m(f, \delta) = 0$. Let $P_i = E^A((t_{i-1}, t_i])$. Then, $\{P_i\}_{i=1}^{n}$ are mutually orthogonal projections whose summation is I. Since P_i commutes with $f(A)$, for every $x \in \mathcal{H}$, we have

$$\|(S(f, \Delta, \{\xi_i\}, E^A) - f(A))x\|^2$$

$$= \left\|\sum_{i=1}^{n}(f(\xi_i)I - f(A))P_ix\right\|^2 = \sum_{i=1}^{n}\|(f(\xi_i)I - f(A))P_ix\|^2$$

$$\leq \sum_{i=1}^{n}\|(f(\xi_i)I - f(A))P_i\|^2\|P_ix\|^2$$

$$\leq \sum_{i=1}^{n}\sup_{t\in\sigma(A)\cap[t_{i-1},t_i]}|f(\xi_i) - f(t)|^2\|P_ix\|^2$$

$$\leq m(f, h(\Delta))^2 \sum_{i=1}^{n}\|P_ix\|^2 = m(f, h(\Delta))^2\|x\|^2,$$

and

$$\|S(f, \Delta, \{\xi_i\}, E^A) - f(A)\| \leq m(f, h(\Delta)) \to 0 \quad (h(\Delta) \to 0)$$

holds.   □

We show the uniqueness of the spectral family satisfying the condition in the above theorem in the following section.

9.2   Borel Functional Calculus

In this section, we construct a normal operator by integrating a bounded Borel function by the spectral family $E = \{E_\lambda\}_{\lambda\in\mathbb{R}}$. Applying this argument to E^A obtained in the previous section, we get the Borel functional calculus for a self-adjoint operator A.

In the following discussion, we fix a spectral family $E = \{E_\lambda\}_{\lambda\in\mathbb{R}}$ acting on a Hilbert space $\mathcal{H}$.

While we used Riemann–Stieltjes integral in the previous section, it is more convenient to construct a (complex) measure and use the integration by the measure. For $x \in \mathcal{H}$, the function $t \mapsto \langle E_tx, x \rangle$ is right continuous, monotone increasing, and satisfies $\lim_{t\to-\infty}\langle E_tx, x \rangle = 0$ and $\lim_{t\to\infty}\langle E_tx, x \rangle = \|x\|^2$. For such a function, there exists a unique bounded Borel measure μ_x on $\mathbb{R}$

satisfying $\mu_x((-\infty, t]) = \langle E_t x, x \rangle$ for all $t \in \mathbb{R}$ (see Section A.6). Also, since

$$\langle E_t x, y \rangle = \frac{1}{4} \sum_{k=0}^{3} i^k \langle E_t(x + i^k y), x + i^k y \rangle,$$

there exists a unique complex measure $\mu_{x,y}$ satisfying $\mu_{x,y}((-\infty, t]) = \langle E_t x, y \rangle$ for all $t \in \mathbb{R}$. The reader need not be afraid of the complex measure because it is concretely expressed as a linear combination of finite measures,

$$\mu_{x,y} = \frac{1}{4} \sum_{k=0}^{3} i^k \mu_{x+i^k y},$$

and we do not need a rather complicated general theory of complex measures. An important fact here is that, since the Borel σ-algebra $\mathfrak{B}_\mathbb{R}$ for $\mathbb{R}$ is generated by the sets of the form $(-\infty, t]$, we have the uniqueness of $\mu_{x,y}$. As a consequence, we see that the map $x \mapsto \mu_{x,y}$ is linear and $y \mapsto \mu_{x,y}$ is conjugate linear.

We say that $\lambda \in \mathbb{R}$ is an increasing point of E if $E_{\lambda+\varepsilon} \neq E_{\lambda-\varepsilon}$ for all $\varepsilon > 0$. We denote by Ω the set of increasing points of E. Then, Ω is a closed subset of $\mathbb{R}$. From the definition of μ_x, we have $\operatorname{supp} \mu_x \subset \Omega$.

We denote by $\mathcal{B}^b(\Omega)$ the set of bounded Borel functions on Ω. The space $\mathcal{B}^b(\Omega)$ is a Banach algebra with the norm $\|\cdot\|_\infty$. We can regard $f \in \mathcal{B}^b(\Omega)$ as an element in $\mathcal{B}^b(\mathbb{R})$ by setting $f(t) = 0$ for $t \in \mathbb{R} \backslash \Omega$.

Lemma 9.4. *For every $f \in \mathcal{B}^b(\Omega)$, there exists a unique operator $\pi_E(f) \in \mathbf{B}(\mathcal{H})$ satisfying the following condition:* $\langle \pi_E(f)x, y \rangle = \int_\Omega f(t) d\mu_{x,y}(t)$ *for all $x, y \in \mathcal{H}$.*

Proof. For $x, y \in \mathcal{H}$, let $b(x, y) = \int_\Omega f(t) d\mu_{x,y}(t)$. Then, b is a sesquilinear form. For $\|x\|, \|y\| \leq 1$, we have

$$|b(x, y)| = \frac{1}{4} \left| \sum_{k=0}^{3} i^k \int_\Omega f(t) d\mu_{x+i^k y}(t) \right| \leq \frac{1}{4} \left| \sum_{k=0}^{3} \int_\Omega |f(t)| d\mu_{x+i^k y}(t) \right|$$

$$\leq \frac{\|f\|_\infty}{4} \sum_{k=0}^{3} \mu_{x+i^k y}(\Omega) = \frac{\|f\|_\infty}{4} \sum_{k=0}^{3} \|x + i^k y\|^2$$

$$= \|f\|_\infty (\|x\|^2 + \|y\|^2) \leq 2\|f\|_\infty,$$

and b is bounded. Thus, there exists a unique bounded operator $\pi_E(f)$ satisfying $\langle \pi_E(f)x, y \rangle = b(x, y)$ for all $x, y \in \mathcal{H}$. □

Lemma 9.5. *For every $f \in \mathcal{B}^b(\Omega)$, we have $d\mu_{\pi_E(f)x,y} = f d\mu_{x,y}$ (that is, $\mu_{\pi_E(f)x,y}(U) = \int_U f(t) d\mu_{x,y}(t)$ holds for every $U \in \mathfrak{B}_\Omega$).*

Proof. It suffices to show $\mu_{\pi_E(f)x,y}((-\infty, \lambda]) = \int_{(-\infty,\lambda]} f(t) d\mu_{x,y}(t)$ for every $\lambda \in \mathbb{R}$. The left-hand side is

$$\langle E_\lambda \pi_E(f)x, y \rangle = \langle \pi_E(f)x, E_\lambda y \rangle = \int_\Omega f(t) d\mu_{x, E_\lambda y}(t).$$

Here, since

$$\mu_{x, E_\lambda y}((-\infty, \mu]) = \langle E_\mu x, E_\lambda y \rangle = \langle E_{\min\{\mu, \lambda\}} x, y \rangle$$
$$= \mu_{x,y}((-\infty, \min\{\mu, \lambda\}])$$
$$= \int_{(-\infty,\mu]} \chi_{(-\infty,\lambda]}(t) d\mu_{x,y}(t)$$

shows that $d\mu_{x, E_\lambda y}(t) = \chi_{(-\infty,\lambda]}(t) d\mu_{x,y}(t)$, we get

$$\int_\Omega f(t) d\mu_{x, E_\lambda y}(t) = \int_\Omega f(t) \chi_{(-\infty,\lambda]}(t) d\mu_{x,y} = \int_{(-\infty,\lambda]} f(t) d\mu_{x,y}(t).$$

□

In a similar way, we have $d\mu_{x, \pi_E(f)y}(t) = \overline{f(t)} d\mu_{x,y}(t)$.

Theorem 9.3. *The map $\pi_E : \mathcal{B}^b(\Omega) \to \mathbf{B}(\mathcal{H})$ is a homomorphism of algebras over $\mathbb{C}$, and $\|\pi_E(f)\| \leq \|f\|_\infty$ and $\pi_E(f)^* = \pi_E(\bar{f})$ hold for every $f \in \mathcal{B}^b(\Omega)$. In particular, the operator $\pi_E(f)$ is normal, and it is self-adjoint if f is real-valued.*

Proof. For $f, g \in \mathcal{B}^b(\Omega)$,

$$\langle \pi_E(f)\pi_E(g)x, y \rangle = \int_\Omega f(t) d\mu_{\pi_E(g)x,y}(t) = \int_\Omega f(t)g(t) d\mu_{x,y}(t)$$
$$= \langle \pi_E(fg)x, y \rangle,$$

and π_E is a homomorphism.

Since

$$\langle x, \pi_E(f)y \rangle = \int_\Omega 1 d\mu_{x,\pi_E(f)y} = \int_\Omega \overline{f(t)} d\mu_{x,y},$$

we get $\pi_E(f)^* = \pi_E(\overline{f})$.

$$\|\pi_E(f)x\|^2 = \langle \pi_E(f)^* \pi_E(f)x, x \rangle = \langle \pi_E(|f|^2)x, x \rangle = \int_\Omega |f(t)|^2 d\mu_x(t)$$

$$\leq \|f\|_\infty^2 \mu_x(\Omega) = \|f\|_\infty^2 \|x\|^2$$

implies $\|\pi_E(f)\| \leq \|f\|_\infty$. $\qquad \square$

We denote $\pi_E(f)$ by $\int_\Omega f(t) dE_t$ or $\int_\mathbb{R} f(t) dE_t$ and call it the *spectral integral*. Since $\{E_t\}_{t \in \mathbb{R}}$ is monotone increasing, the left limit s-$\lim_{t \to \lambda - 0} E_t = E_{\lambda - 0}$ exists. Since $E_t \leq E_\lambda$ for every $t < \lambda$, we have $E_{\lambda - 0} \leq E_\lambda$.

Proposition 9.1. *Assume that Ω is a bounded set, and let $A = \int_\Omega t dE_t \in \mathbf{B}(\mathcal{H})_{\mathrm{sa}}$. Then, the following hold:*

(1) $\sigma(A) = \Omega$.
(2) *The operator $E_\lambda - E_{\lambda - 0}$ is the projection onto $\ker(\lambda I - A)$ for every $\lambda \in \mathbb{R}$. In particular, the following holds: $\lambda \in \sigma_{\mathrm{p}}(A) \iff E_\lambda \neq E_{\lambda - 0}$.*
(3) *If $f \in C(\sigma(A))$, we have $\pi_E(f) = f(A)$.*

Proof. (1) For $\lambda \in \mathbb{C} \setminus \Omega$, since $g(t) = (\lambda - t)^{-1}$ is bounded on Ω, we have $\pi_E(g)(\lambda I - A) = (\lambda I - A)\pi_E(g) = I$, and $\lambda \in \rho(A)$. For $\lambda \in \Omega$, we have $E((\lambda - 1/n, \lambda + 1/n]) \neq 0$ for every $n \in \mathbb{N}$, and there exists $x_n \in \mathcal{R}(E((\lambda - 1/n, \lambda + 1/n]))$ with $\|x_n\| = 1$. Then,

$$\|(A - \lambda I)x_n\|^2 = \int_{(\lambda - \frac{1}{n}, \lambda + \frac{1}{n}]} |t - \lambda|^2 d\mu_{x_n}(t)$$

$$\leq \frac{1}{n^2} \mu_{x_n}\left(\left(\lambda - \frac{1}{n}, \lambda + \frac{1}{n}\right]\right)$$

$$= \frac{1}{n^2} \to 0 \quad (n \to \infty),$$

and $\lambda \in \sigma_{\mathrm{ap}}(A)$.

(2) Since $\|Ax - \lambda x\|^2 = \int_\Omega |t - \lambda|^2 d\mu_x(t)$, we get

$$x \in \ker(\lambda I - A) \iff \operatorname{supp} \mu_x \subset \{\lambda\} \iff x \in \mathcal{R}(E_\lambda - E_{\lambda-0}).$$

(3) Since $\pi_E : \mathcal{B}^b(\Omega) \to \mathbf{B}(\mathcal{H})$ is a homomorphism of algebras over $\mathbb{C}$, and $\pi_E(1) = I$, and $\pi_E(t) = A$, we have $\pi_E(f) = f(A)$ for every polynomial $f(t)$. Further, $\|\pi_E(f)\| \leq \|f\|_\infty$ implies $\pi_E(f) = f(A)$ for every $f \in C(\sigma(A))$. $\qquad\square$

Theorem 9.4 (Uniqueness of spectral decomposition). *Assume that Ω is a bounded set, and let $A = \int_\Omega t\,dE_t \in \mathbf{B}(\mathcal{H})_{\mathrm{sa}}$. Then, $E^A = E$.*

Proof. Let $\{f_{\lambda,n}\}_{n=1}^\infty$ be the sequence of functions defined in the previous section. Then, we have $E_\lambda^A = \text{s-}\lim_{n\to\infty} f_{\lambda,n}(A)$. On the other hand, the bounded convergence theorem shows that

$$\lim_{n\to\infty} \langle f_{\lambda,n}(A)x, x\rangle = \lim_{n\to\infty} \int_\Omega f_{\lambda,n}(t)d\mu_x(t) = \int_\Omega \chi_{(-\infty,\lambda]}(t)d\mu_x(t)$$
$$= \langle E_\lambda x, x\rangle.$$

Thus, $E_\lambda^A = E_\lambda$. $\qquad\square$

When $A = \int_\Omega t\,dE_t$, we denote $\pi_E(f)$ by $f(A)$.

For a unitary operator $U \in \mathcal{U}(\mathcal{H})$ and a closed subset F of $\sigma(U)$, we can define a projection $E^U(F)$ in exactly the same way as in the case of self-adjoint operators. For $0 \leq \lambda < 2\pi$, we set $F_\lambda = \{e^{it} \in \mathbb{C}; t \in [0, \lambda]\} \cap \sigma(U)$, and we set

$$E_t^U = \begin{cases} 0, & t < 0, \\ E^U(F_t), & 0 \leq t < 2\pi, \\ I, & 2\pi \leq t. \end{cases}$$

Then, $E^U = \{E_t^U\}_{t\in\mathbb{R}}$ is a spectral family.

Lemma 9.6. *We have $E_{2\pi-0}^U = I$.*

Proof. For $\pi < \lambda < 2\pi$, in the same way as in Lemma 9.3, we can show that

$$\|(I - U)(I - E_\lambda^U)\| \leq \sup_{\lambda \leq t \leq 2\pi} |1 - e^{it}| = 2\sin\frac{\lambda}{2}.$$

Assume that $x \in \mathcal{H}$ satisfies $E^U_{2\pi-0}x = 0$. Since $E^U_\lambda x = 0$ for every $\pi < \lambda < 2\pi$, we get

$$\|(I - U)x\| = \|(I - U)(I - E^U_\lambda)x\| \le 2\sin\frac{\lambda}{2}\|x\|.$$

Since this is the case for every $\pi < \lambda < 2\pi$, we get $x \in \ker(I - U)$. Theorem 6.6(3) (in the case of unitary operators) shows $E^U_0 x = x$, and $x = 0$. Thus, $E^U_{2\pi-0} = I$. $\square$

Our arguments so far show the following.

Theorem 9.5. *For every $U \in \mathcal{U}(\mathcal{H})$, there exists a unique spectral family $E^U = \{E^U_t\}_{t\in\mathbb{R}}$ satisfying the following conditions:*

(1) $U = \int_{\mathbb{R}} e^{it} dE^U_t$.
(2) $E^U_t = 0$ *for every* $t < 0$, *and* $E^U_{2\pi-0} = I$.

Moreover, there exists a unique $A \in \mathbf{B}(\mathcal{H})_{\mathrm{sa}}$ satisfying the following conditions:

(1) $U = e^{iA}$.
(2) $0 \le A \le 2\pi I$ *and* $2\pi \notin \sigma_{\mathrm{p}}(A)$.

Exercises

Exercise 9.1
Show Exercise 4.5(2) and Problem 9.1 by using spectral decomposition.

Exercise 9.2
Let $A = \int_{\sigma(A)} t dE^A_t$ be the spectral decomposition of $A \in \mathbf{B}(\mathcal{H})_{\mathrm{sa}}$. Show the following:

(1) For $x \in \mathcal{H}$, the measure $d\langle E^A_t x, x\rangle$ is uniquely determined by the moment series $\{\langle A^n x, x\rangle\}_{n=0}^{\infty}$.
(2) Let U be the bilateral shift of $\ell^2(\mathbb{Z})$, and let $A = 2^{-1}(U + U^{-1})$. Then,

$$\langle A^n \delta_0, \delta_0\rangle = \frac{1}{\pi}\int_{-1}^{1}\frac{t^n}{\sqrt{1 - t^2}}dt.$$

(3) Let V be the unilateral shift of ℓ^2, and let $A = 2^{-1}(V + V^*)$. Then,

$$\langle A^n \delta_1, \delta_1 \rangle = \frac{2}{\pi} \int_{-1}^{1} t^n \sqrt{1 - t^2} dt.$$

Exercise 9.3

We use the notation in Exercise 7.6 for this exercise. For $A \in \mathbf{B}(\mathcal{H})_{\mathrm{sa}}$, we denote by $\Sigma(A)$ the union of the set of accumulation points of $\sigma(A)$ and the set of eigenvalues of A with infinite multiplicity. Show $\sigma_{\mathrm{e}}(A) = \sigma_{\mathrm{w}}(A) = \Sigma(A)$ by showing the following statements:

(1) For $\lambda \in \Sigma(A)$, there exists an ONS $\{e_n\}_{n=1}^{\infty}$ in $\mathcal{H}$ satisfying $\lim_{n \to \infty} \|A e_n - \lambda e_n\| = 0$.
(2) $\sigma_{\mathrm{w}}(A) \supset \Sigma(A)$.
(3) If $\lambda \in \sigma(A)$ is an isolated point and $\dim(\lambda I - A) < \infty$, we have $\lambda \notin \sigma_{\mathrm{e}}(A)$.

Exercise 9.4

Assume that $\mathcal{H}$ is separable, and let $A \in \mathbf{B}(\mathcal{H})_{\mathrm{sa}}$. For $x \in \mathcal{H}$, let $\mathcal{H}_{A,x} = \overline{\mathrm{span}\{A^n x\}_{n=0}^{\infty}}$.

(1) Assume that there exists $x_0 \in \mathcal{H}$ satisfying $\mathcal{H}_{A,x_0} = \mathcal{H}$. Show that there exists a finite measure μ on $\sigma(A)$ such that A is unitarily equivalent to the multiplication operator $M_t \in \mathbf{B}(L^2(\sigma(A), \mu))$ of the coordinate function t. We say that A satisfying this condition has a *simple spectrum*.
(2) Show that, in general, there exists a measure space (Ω, μ) and bounded measurable function f such that A is unitarily equivalent to the multiplication operator $M_f \in \mathbf{B}(L^2(\Omega, \mu))$ of f.

Exercise 9.5

Let $c_{00}(\mathbb{Z})$ be the set of complex-valued sequences $(x_n)_{n \in \mathbb{Z}}$ satisfying $x_n = 0$ except for finitely many n. We say that a sequence $(a_n)_{n \in \mathbb{Z}}$ is positive definite if $\sum_{m,n \in \mathbb{Z}} a_{m-n} x_m \overline{x_n} \geq 0$ holds for every $x \in c_{00}(\mathbb{Z})$. By showing (1), (2), and (3), as follows, using the spectral decomposition of a unitary operator, show the following Herglotz theorem: For every positive definite sequence $(a_n)_{n \in \mathbb{Z}}$, there exists a finite measure μ on $[0, 2\pi)$ such that $a_n = \int_{[0,2\pi)} e^{int} d\mu(t)$ holds for every $n \in \mathbb{Z}$.

(1) Let N be the set of x in $c_{00}(\mathbb{Z})$ satisfying $\sum_{m,n\in\mathbb{Z}} a_{m-n} x_m \overline{x_n} = 0$. Show that N is a linear subspace of $c_{00}(\mathbb{Z})$ and that $\langle [x], [y] \rangle = \sum_{m,n\in\mathbb{Z}} a_{m-n} x_m \overline{y_n}$ is an inner product of $c_{00}(\mathbb{Z})/N$.

(2) Let $(U_0 x)_n = x_{n-1}$. Let $\mathcal{H}$ be the completion of $c_{00}(\mathbb{Z})/N$ with respect to the norm given by the above inner product. Show that there exists a unitary operator U on $\mathcal{H}$ satisfying $U[x] = [U_0 x]$ for all $x \in c_{00}(\mathbb{Z})$.

(3) Using $a_n = \langle U^n[\delta_0], [\delta_0] \rangle$, show the Herglotz theorem.

Chapter 10

Unbounded Operators
on Hilbert Spaces

The goal of the final chapter is to prove the spectral decomposition theorem for unbounded self-adjoint operators. It is impossible to overemphasize the importance of the domain of an operator in the theory of unbounded operators. We see, in the simplest example of the differential operator in dimension 1, that self-adjoint extensions of a symmetric operator correspond to boundary conditions.

10.1 Closed Operators on Hilbert Spaces

Let $\mathcal{H}$, $\mathcal{H}_i$, $i = 1, 2, \ldots$, be Hilbert spaces throughout this section. Although we have already introduced an operator with a domain in Chapter 4, here for the purpose of review and clarifying notation, we begin our discussion by introducing definitions and notation again.

In what follows, by an operator T from $\mathcal{H}_1$ to $\mathcal{H}_2$, we mean a linear map from a subspace $\mathcal{D}(T)$ of $\mathcal{H}_1$ to $\mathcal{H}_2$, and we call $\mathcal{D}(T)$ the *domain* of T. If two operators T_1 and T_2 from $\mathcal{H}_1$ to $\mathcal{H}_2$ satisfy $\mathcal{D}(T_1) \subset \mathcal{D}(T_2)$ and $T_2|_{\mathcal{D}(T_1)} = T_1$, we say that T_2 is an *extension* of T_1 and write $T_1 \subset T_2$.

For operators S, T from $\mathcal{H}_1$ to $\mathcal{H}_2$, we define $S + T$ by setting $\mathcal{D}(S + T) = \mathcal{D}(S) \cap \mathcal{D}(T)$ and $(S + T)x = Sx + Tx$. If $\alpha \in \mathbb{C}$, we define αT by setting $\mathcal{D}(\alpha T) = \mathcal{D}(T)$ and $(\alpha T)x = \alpha(Tx)$.

When T is an operator from $\mathcal{H}_1$ to $\mathcal{H}_2$ and S is an operator from $\mathcal{H}_2$ to $\mathcal{H}_3$, we define the operator ST from $\mathcal{H}_1$ to $\mathcal{H}_3$ by setting $\mathcal{D}(ST) = \{x \in \mathcal{D}(T); Tx \in \mathcal{D}(S)\}$ and $(ST)x = S(Tx)$.

Remark 10.1. Even when $\mathcal{D}(S)$ and $\mathcal{D}(T)$ are dense subspaces in $\mathcal{H}_1$ or $\mathcal{H}_2$ in the above argument, it could happen that $\mathcal{D}(S + T)$ or $\mathcal{D}(ST)$ is $\{0\}$.

Let S be an operator on $\mathcal{H}_1$, and let T be an operator on $\mathcal{H}_2$. We say that S and T are *unitarily equivalent* if there exists a unitary $U \in \mathcal{U}(\mathcal{H}_1, \mathcal{H}_2)$ satisfying $U\mathcal{D}(S) = \mathcal{D}(T)$ and $US = TU|_{\mathcal{D}(S)}$.

Example 10.1. Let $\mathcal{H} = L^2(\mathbb{R})$.

(1) Let $\mathcal{D}(A_0)$ be the Schwartz space $\mathcal{S}(\mathbb{R})$ (see Example 5.3), and let $A_0 f(t) = tf(t)$. Let $\mathcal{D}(A) = \{f \in L^2(\mathbb{R}); \int_{\mathbb{R}} t^2 |f(t)|^2 dt < \infty\}$, and let $Af(t) = tf(t)$. Then, we have $A_0 \subset A$.

(2) Let $\mathcal{D}(B_0) = \mathcal{S}(\mathbb{R})$, and let $B_0 f(t) = -if'(t)$. Let $\mathcal{F}$ be the Fourier transform of $L^2(\mathbb{R})$. Then, $\mathcal{F}$ maps $\mathcal{S}(\mathbb{R})$ to itself bijectively, and $\mathcal{F}f(\xi) = (2\pi)^{-1/2} \int_{\mathbb{R}} f(t)e^{-it\xi} dt$ holds for $f \in \mathcal{S}(\mathbb{R})$. Since $\mathcal{F}B_0 = A_0 \mathcal{F}|_{\mathcal{S}(\mathbb{R})}$ holds, the two operators A_0 and B_0 are unitarily equivalent.

In what follows, we denote $\mathcal{H}_1 \oplus_2 \mathcal{H}_2$ by $\mathcal{H}_1 \oplus \mathcal{H}_2$ for simplicity. Then, $\mathcal{H}_1 \oplus \mathcal{H}_2$ is a Hilbert space with the inner product

$$\langle (x_1, x_2), (y_1, y_2) \rangle = \langle x_1, y_1 \rangle + \langle x_2, y_2 \rangle.$$

Definition 10.1. We define the graph of an operator from $\mathcal{H}_1$ to $\mathcal{H}_2$ by

$$\mathcal{G}(T) = \{(x, Tx) \in \mathcal{H}_1 \oplus \mathcal{H}_2; x \in \mathcal{D}(T)\}.$$

We say that T is a *closed operator* if $\mathcal{G}(T)$ is a closed subspace of $\mathcal{H}_1 \oplus \mathcal{H}_2$. We say that T is *closable* if T has an extension that is a closed operator.

The condition that T is a closed operator is equivalent to the following: If a sequence $\{x_n\}_{n=1}^{\infty}$ in $\mathcal{D}(T)$ converges to $x \in \mathcal{H}_1$ and $\{Tx_n\}_{n=1}^{\infty}$ converges to $y \in \mathcal{H}_2$, then $x \in \mathcal{D}(\mathcal{H}_1)$ and $y = Tx$ hold.

When T is a closed operator, the graph $\mathcal{G}(T)$ is a closed subspace of $\mathcal{H}_1 \oplus \mathcal{H}_2$, and it is a Hilbert space with the inner product $\langle (x, Tx), (y, Ty) \rangle = \langle x, y \rangle + \langle Tx, Ty \rangle$. Thus, if we define the *graph inner product* for $x, y \in \mathcal{D}(T)$ by $\langle x, y \rangle_T = \langle x, y \rangle + \langle Tx, Ty \rangle$, then $(\mathcal{D}(T), \langle \cdot, \cdot \rangle_T)$ is a Hilbert space, and the map $\mathcal{D}(T) \to \mathcal{G}(T)$, $x \mapsto (x, Tx)$, is a unitary operator.

If T is a closed operator and $S \in \mathbf{B}(\mathcal{H}_1, \mathcal{H}_2)$, then $S + T$ is a closed operator.

The following easy lemma is useful for characterizing the condition for an operator to be closable.

Lemma 10.1. *For a subspace $\mathcal{K}$ of $\mathcal{H}_1 \oplus \mathcal{H}_2$, the following conditions are equivalent:*

(1) *There exists an operator from $\mathcal{H}_1$ to $\mathcal{H}_2$ whose graph is $\mathcal{K}$.*
(2) $\mathcal{K} \cap (\{0\} \oplus \mathcal{H}_2) = \{0\}$.

Proof. (1) $\Longrightarrow$ (2) follows from the definition of the graph of an operator.

(2) $\Longrightarrow$ (1). Let $\mathcal{D}(T) = \{x \in \mathcal{H}_1;\ \exists y \in \mathcal{H}_2,\ (x, y) \in \mathcal{K}\}$. Then, $\mathcal{D}(T)$ is a subspace of $\mathcal{H}_1$ because $\mathcal{K}$ is a subspace of $\mathcal{H}_1 \oplus \mathcal{H}_2$. For a fixed $x \in \mathcal{D}(T)$, there exists a unique $y \in \mathcal{H}_2$, with $(x, y) \in \mathcal{K}$. Indeed, if $y' \in \mathcal{H}_2$ satisfies the same condition, then $(x, y) - (x, y') = (0, y - y') \in \mathcal{K}$, and $y = y'$. From the uniqueness, we can introduce a map $T : \mathcal{D}(T) \to \mathcal{H}_2$ assigning y to $x \in \mathcal{D}(T)$ if $(x, y) \in \mathcal{K}$. Since $\mathcal{K}$ is a subspace, the map T is linear. By the construction of T, we get $\mathcal{K} = \mathcal{G}(T)$. $\qquad\square$

When an operator T from $\mathcal{H}_1$ to $\mathcal{H}_2$ is closable, by definition there exists a closed extension $T \subset T_1$. Then, we have $\mathcal{G}(T) \subset \overline{\mathcal{G}(T)} \subset \mathcal{G}(T_1)$, and

$$\overline{\mathcal{G}(T)} \cap (\{0\} \oplus \mathcal{H}_2) \subset \mathcal{G}(T_1) \cap (\{0\} \oplus \mathcal{H}_2) = \{0\}.$$

Thus, the above lemma shows that there exists an operator $\overline{T}$ satisfying $\overline{\mathcal{G}(T)} = \mathcal{G}(\overline{T})$. We call $\overline{T}$ the *closure* of T. The operator $\overline{T}$ is the smallest closed extension of T.

Since T is closable if and only if $\overline{\mathcal{G}(T)} \cap (\{0\} \oplus \mathcal{H}_2) = \{0\}$, we see the following.

Lemma 10.2. *The condition that T is closable is equivalent to the following: If $\{x_n\}_{n=1}^{\infty}$ is a sequence in $\mathcal{D}(T)$ converging to 0 and $\{Tx_n\}_{n=1}^{\infty}$ converges to $y \in \mathcal{H}_2$, then $y = 0$.*

Next, we introduce the notion of the adjoint operator of an unbounded operator. Its definition is subtle compared to the case of bounded operators, and we need to handle it carefully.

Definition 10.2. Let T be an operator from $\mathcal{H}_1$ to $\mathcal{H}_2$, and assume that $\mathcal{D}(T)$ is dense in $\mathcal{H}_1$. We set

$$\mathcal{D}(T^*) = \{y \in \mathcal{H}_2;\; \mathcal{D}(T) \ni x \mapsto \langle Tx, y \rangle \in \mathbb{C} \text{ is bounded}\}.$$

When $y \in \mathcal{D}(T^*)$, since $\mathcal{D}(T)$ is dense in $\mathcal{H}_1$, the linear functional $x \mapsto \langle Tx, y \rangle$ uniquely extends to a bounded linear functional on $\mathcal{H}_1$, and the Riesz representation theorem shows that there exists a unique $y^* \in \mathcal{H}_1$ satisfying $\langle Tx, y \rangle = \langle x, y^* \rangle$ for every $x \in \mathcal{D}(T)$. We denote y^* by T^*y and call T^* the *adjoint operator* of T. The map $T^* : \mathcal{D}(T^*) \to \mathcal{H}_1$ is linear. Alternatively, we could define $\mathcal{D}(T^*)$ by

$$\{y \in \mathcal{H}_2;\; \exists y^* \in \mathcal{H}_1,\; \forall x \in \mathcal{D}(T),\; \langle Tx, y \rangle = \langle x, y^* \rangle\}.$$

As in the case of bounded operators, we have $\mathcal{R}(T)^\perp = \ker T^*$. Also, if $S \subset T$, we have $T^* \subset S^*$.

We may say that the following theorem is the only basis for analyzing unbounded operators on Hilbert spaces. We define $\mathcal{V} \in \mathcal{U}(\mathcal{H}_1 \oplus \mathcal{H}_2, \mathcal{H}_2 \oplus \mathcal{H}_1)$ by $\mathcal{V}(x, y) = (-y, x)$.

Theorem 10.1. *Let T be an operator from $\mathcal{H}_1$ to $\mathcal{H}_2$, and assume that $\mathcal{D}(T)$ is dense in $\mathcal{H}_1$. Then, the following hold:*

(1) $\mathcal{G}(T^*) = (\mathcal{V}\mathcal{G}(T))^\perp = \mathcal{V}(\mathcal{G}(T)^\perp)$.
(2) T *is closable* $\iff \mathcal{D}(T^*)$ *is dense in* $\mathcal{H}_2$.

Proof. (1) For $(y, z) \in \mathcal{H}_2 \oplus \mathcal{H}_1$, we have

$$(y, z) \in \mathcal{V}(\mathcal{G}(T)^\perp) \iff (z, -y) \in \mathcal{G}(T)^\perp$$
$$\iff \forall x \in \mathcal{D}(T),\; (x, Tx) \perp (z, -y)$$
$$\iff \forall x \in \mathcal{D}(T),\; \langle Tx, y \rangle = \langle x, z \rangle.$$

Since this condition is equivalent to $y \in \mathcal{D}(T^*)$ and $T^*y = z$, we get $\mathcal{V}(\mathcal{G}(T)^\perp) = \mathcal{G}(T^*)$.

(2) We can compute $\overline{\mathcal{G}(T)} \cap (\{0\} \oplus \mathcal{H}_2)$ as

$$\overline{\mathcal{G}(T)} \cap (\{0\} \oplus \mathcal{H}_2) = \mathcal{G}(T)^{\perp\perp} \cap (\{0\} \oplus \mathcal{H}_2)$$
$$= \mathcal{V}^{-1}((\mathcal{V}(\mathcal{G}(T)^{\perp}))^{\perp}) \cap (\{0\} \oplus \mathcal{H}_2)$$
$$= \mathcal{V}^{-1}(\mathcal{G}(T^*)^{\perp} \cap (\mathcal{H}_2 \oplus \{0\})).$$

Here, the right-hand side is $\mathcal{V}^{-1}(\mathcal{D}(T^*)^{\perp} \oplus \{0\}) = \{0\} \oplus \mathcal{D}(T^*)^{\perp}$, and we get the statement. $\qquad\square$

Corollary 10.1. *Under the assumption of the above theorem, the following hold:*

(1) T^* *is a closed operator.*
(2) *If T is closable, we have $T^{**} = \overline{T}$ and $T^* = T^{***}$. In particular, $T = T^{**}$ if T is a closed operator.*

Proof. (1) Since $\mathcal{G}(T^*) = \mathcal{V}(\mathcal{G}(T)^{\perp})$, the operator T^* is closed.

(2) When T is closable, since $\mathcal{D}(T^*)$ is dense in $\mathcal{H}_2$, the operator T^{**} is defined, and the computation in the above proof shows $\overline{\mathcal{G}(T)} = \mathcal{V}^{-1}(\mathcal{G}(T^*)^{\perp}) = \mathcal{G}(T^{**})$. Thus, $\overline{T} = T^{**}$. Since T^* is a closed operator, we get $T^* = T^{***}$. $\qquad\square$

Definition 10.3. Let T be an operator from $\mathcal{H}_1$ to $\mathcal{H}_2$, and assume that $\mathcal{D}(T)$ is dense in $\mathcal{H}_1$. We say that T is *invertible* and write $S = T^{-1}$ if there exists $S \in \mathbf{B}(\mathcal{H}_2, \mathcal{H}_1)$ satisfying $ST \subset I_{\mathcal{H}_1}$ and $TS = I_{\mathcal{H}_2}$. Since $\mathcal{G}(T) = \{(T^{-1}y, y) \in \mathcal{H}_1 \oplus \mathcal{H}_2; \ y \in \mathcal{H}_2\}$, such an operator T is automatically closed.

The closed graph theorem implies the following.

Proposition 10.1. *Let T be an operator from $\mathcal{H}_1$ to $\mathcal{H}_2$, and assume that $\mathcal{D}(T)$ is dense in $\mathcal{H}_1$. Then, the following conditions are equivalent:*

(1) T *is invertible.*
(2) T *is a closed operator, and $T : \mathcal{D}(T) \to \mathcal{H}_2$ is a bijection.*

Definition 10.4. Let T be a closed operator on $\mathcal{H}$, and assume that $\mathcal{D}(T)$ is dense in $\mathcal{H}$. We define the spectrum $\sigma(T)$ and the resolvent set $\rho(T)$ of T, and further $\sigma_{\mathrm{p}}(T)$, $\sigma_{\mathrm{c}}(T)$, $\sigma_{\mathrm{r}}(T)$, and $\sigma_{\mathrm{ap}}(T)$, as in the case where T is bounded. The set $\sigma(T)$ is always closed but is not necessarily compact, and it could be empty (see Exercise 10.1). Since T is a closed operator, the space $\ker(\lambda I - T)$ is closed for $\lambda \in \sigma_{\mathrm{p}}(T)$.

Definition 10.5. Let S be an operator on $\mathcal{H}$ with dense domain.

- We say that S is a *symmetric operator* if $S \subset S^*$ holds. This condition is equivalent to the one that $\langle Sx, y \rangle = \langle x, Sy \rangle$ holds for all $x, y \in \mathcal{D}(S)$. Every symmetric operator S is closable because S^* is a closed operator. We have $S \subset S^{**} = \overline{S} \subset S^*$.
- If a symmetric operator S satisfies $S = S^*$, we say that S is *self-adjoint.*
- If the closure of a symmetric operator S is self-adjoint, we say that S is *essentially self-adjoint.* This condition is equivalent to $S^* = S^{**}$.

10.2 Concrete Examples of Symmetric Operators

Unlike the case of bounded operators, the difference between being symmetric and being self-adjoint for unbounded operators is huge and requires careful treatment. To understand this, we take a look at fundamental examples in this section before developing a general theory.

We start with the easiest example to handle. The operators A_0, A, and B_0 in Example 10.1 are examples of symmetric operators. We show that A_0 and B_0 are essentially self-adjoint and A is self-adjoint. First, we show $A_0{}^* = A$. For every $f \in \mathcal{D}(A_0)$ and every $g \in \mathcal{D}(A)$, we have $\langle A_0 f, g \rangle = \langle f, Ag \rangle = \int_{\mathbb{R}} t f(t) \overline{g(t)} dt$, and $A \subset A_0{}^*$ holds. Conversely, let $g \in \mathcal{D}(A_0{}^*)$, and let $g^* = A_0{}^* g$. Then, for every $f \in \mathcal{D}(A_0) = \mathcal{S}(\mathbb{R})$, we have $\int_{\mathbb{R}} t f(t) \overline{g(t)} dt = \int_{\mathbb{R}} f(t) \overline{g^*(t)} dt$, and $g^*(t) = t g(t)$ holds for almost every $t \in \mathbb{R}$. This shows $g \in \mathcal{D}(A)$ and $g^* = Ag$, and so $A_0{}^* \subset A$.

Since $A_0{}^*$ is a closed operator, we have $\overline{A_0} \subset A_0{}^*$. On the other hand, taking the adjoint of $A_0{}^* = A$, we get $\overline{A_0} = A^* \supset A = A_0{}^*$, and $\overline{A_0} = A_0{}^* = A$ is self-adjoint. This shows that A_0 is essentially self-adjoint, and so is B_0 because it is unitarily equivalent to A_0.

Next, we see a much subtler example.

Example 10.2. We define an operator D on $\mathcal{H} = L^2(0, 1)$ by $\mathcal{D}(D) = C_c^\infty(0, 1)$ and $Df(t) = -if'(t)$. Then, for all $f, g \in \mathcal{D}(D)$, we have

$$\langle Df, g \rangle - \langle f, Dg \rangle = -i \int_0^1 (f'(t)\overline{g(t)} + f(t)\overline{g'(t)})dt = -if(t)\overline{g(t)}|_0^1 = 0,$$

and D is a symmetric operator.

To determine the adjoint operator of D, we show the following lemma well-known in the theory of distributions.

Lemma 10.3. *Let $n \in \mathbb{N}$. If a linear functional $\varphi : C_c^\infty(0,1) \to \mathbb{C}$ satisfies $\varphi(f^{(n)}) = 0$ for every $f \in C_c^\infty(0,1)$, then there exists a polynomial h of degree at most $n-1$ satisfying $\varphi(f) = \int_0^1 f(t)h(t)dt$ for every $f \in C_c^\infty(0,1)$.*

Proof. We first show the statement for $n = 1$. For $f \in C_c^\infty(0,1)$, let $\omega(f) = \int_0^1 f(t)dt$. We choose and fix $g \in C_c^\infty(0,1)$, with $\int_0^1 g(t)dt = 1$. For a given $f \in C_c^\infty(0,1)$, we set $F(t) = \int_0^t (f(s) - \omega(f)g(s))ds$. Then, $F \in C_c^\infty(0,1)$, and $F'(t) = f(t) - \omega(f)g(t)$ holds. Thus, $0 = \varphi(F') = \varphi(f) - \omega(f)\varphi(g)$, and $\varphi(f) = \varphi(g)\int_0^1 f(t)dt$ holds.

Assume that the statement is true for n and a linear functional $\varphi : C_c^\infty(0,1) \to \mathbb{C}$ satisfies $\varphi(f^{(n+1)}) = 0$ for all $f \in C_c^\infty(0,1)$. Let $\varphi'(f) = -\varphi(f')$. Then, for every $f \in C_c^\infty(0,1)$, we have $\varphi'(f^{(n)}) = 0$, and the induction hypothesis implies that there exists a polynomial $h(t)$ of degree at most $n-1$ satisfying $\varphi'(f) = \int_0^1 f(t)h(t)dt$ for all $f \in C_c^\infty(0,1)$. Let $H(t) = \int_0^t h(s)ds$, and let $\psi(f) = \varphi(f) - \int_0^1 f(t)H(t)dt$. Then,

$$\psi(f') = -\varphi'(f) - \int_0^1 f'(t)H(t)dt = -\int_0^1 (f(t)H(t))'dt = 0$$

holds for all $f \in C_c^\infty(0,1)$. Now, the statement for $n = 1$ implies that there exists a constant $c \in \mathbb{C}$ satisfying $\varphi(f) = \int_0^1 f(t)(H(t) + c)dt$. $\qquad\square$

Using the fact that $C_c^\infty(0,1)$ is dense in $L^2(0,1)$, we can see that if $g \in L^2(0,1)$ satisfies $\langle g, f^{(n)} \rangle = 0$ for every $f \in C_c^\infty(0,1)$, the function g is a polynomial of degree at most $n-1$.

We denote by $AC[0,1]$ the set of absolutely continuous functions on $[0,1]$ (see Section A.7).

Theorem 10.2. *For the operator D on $L^2(0,1)$ defined by $\mathcal{D}(D) = C_c^\infty(0,1)$, $Df = -if'$, the following hold:*

(1) $\mathcal{D}(D^*) = \{f \in \mathrm{AC}[0,1]; \ f' \in L^2(0,1)\}$, $D^*f = -if'$.
(2) $\mathcal{D}(D^{**}) = \{f \in \mathcal{D}(D^*); \ f(0) = f(1) = 0\}$, $D^{**}f = -if'$.

Proof. (1) Let $f \in \mathcal{D}(D)$. If $g \in \mathrm{AC}[0,1]$ satisfies $g' \in L^2(0,1)$, we have

$$\langle Df, g \rangle - \langle f, -ig' \rangle = -if(t)\overline{g(t)}|_0^1 = 0.$$

Thus, $g \in \mathcal{D}(D^*)$ and $D^*g = -ig'$.

On the other hand, for $g \in \mathcal{D}(D^*)$, let $g^* = D^*g$, and let $h(t) = \int_0^t g^*(s)ds$. Then, $h \in \mathrm{AC}[0,1]$, and

$$\langle f, D^*g \rangle = \langle f, h' \rangle = f(t)\overline{h(t)}|_0^1 - \int_0^1 f'(t)\overline{h(t)}dt = -\int_0^1 f'(t)\overline{h(t)}dt$$

holds. Thus,

$$0 = \langle Df, g \rangle - \langle f, D^*g \rangle = \int_0^1 f'(t)\overline{(ig(t) + h(t))}dt,$$

and $ig + h$ is a constant. This shows $g \in \mathrm{AC}[0,1]$ and $D^*g = -ig'$.

(2) Note that we have $D \subset D^{**} \subset D^*$. The above computation with $f, g \in \mathcal{D}(D^*)$ implies

$$\langle D^*f, g \rangle - \langle f, D^*g \rangle = -i(f(1)\overline{g(1)} - f(0)\overline{g(0)}).$$

This shows that $g \in \mathcal{D}(D^{**})$ if and only if $f(0)\overline{g(0)} = f(1)\overline{g(1)}$ for all $f \in \mathcal{D}(D^*)$. In particular, $f(t) = t$ and $f(t) = 1 - t$ imply that $g(0) = g(1) = 0$ is a necessary condition. Conversely, we can see that $g(0) = g(1) = 0$ implies $g \in \mathcal{D}(D^{**})$. $\qquad\square$

From the above theorem, we see that $\overline{D} = D^{**}$ is not self-adjoint. We leave it as a problem to completely determine its self-adjoint extensions.

Problem 10.1. For $c \in \mathbb{C} \cup \{\infty\}$, let

$$\mathcal{D}(D_c) = \{f \in \mathcal{D}(D^*); f(1) = cf(0)\},$$

and let $D_c = D^*|_{\mathcal{D}(D_c)}$. Here, for $c = \infty$, let

$$\mathcal{D}(D_c) = \{f \in \mathcal{D}(D^*); f(0) = 0\}.$$

(1) Show $D_c{}^* = D_{\bar{c}^{-1}}$. In particular, show that D_c is self-adjoint if $|c| = 1$.
(2) Show that for every operator $D^{**} \subsetneq T \subsetneq D^*$, there exists $c \in \mathbb{C} \cup \{\infty\}$ satisfying $T = D_c$. In particular, show that D_c, $|c| = 1$, exhaust all the self-adjoint extensions of D.

Example 10.3. We define a symmetric operator $\tilde{D}$ on $\mathcal{H} = L^2(0, \infty)$ by $\mathcal{D}(\tilde{D}) = C_c^\infty(0, \infty)$ and $\tilde{D}f = -if'$. Applying a similar argument as in the above proof to $(0, R)$, $R > 0$, we can determine $\tilde{D}^*$ as follows. The domain $\mathcal{D}(\tilde{D}^*)$ is the set of functions f whose restriction to every finite closed interval in $[0, \infty)$ is absolutely continuous and $f, f' \in L^2(0, \infty)$, and $\tilde{D}^*f = -if'$. Also, $\mathcal{D}(\tilde{D}^{**}) = \{f \in \mathcal{D}(\tilde{D}^*); f(0) = 0\}$ and $\tilde{D}^{**}f = -if'$ hold. For every $f \in \mathcal{D}(\tilde{D}^*)$, we have $f(t) = (f(t) - f(0)e^{-t}) + f(0)e^{-t}$, and $\mathcal{D}(\tilde{D}^*) = \mathcal{D}(\tilde{D}^{**}) \oplus \mathbb{C}e^{-t}$, which shows $\dim \mathcal{D}(\tilde{D}^*)/\mathcal{D}(\tilde{D}^{**}) = 1$. Thus, an operator A with $\tilde{D}^{**} \subset A \subset \tilde{D}^*$ is either $\tilde{D}^{**}$ or $\tilde{D}^*$, and $\tilde{D}$ has no self-adjoint extension.

So far, we have considered only the first-order differential operator $-i\frac{d}{dt}$. Hereafter, we consider the second-order differential operator restricting ourselves to the case of the interval $(0, 1)$. In what follows, we use the notation in Problem 10.1. It is a fundamental problem to consider the domain on which the differential operator $-\frac{d^2}{dt^2}$ is self-adjoint. Of particular importance among such operators are the Laplacian $-\Delta_D = D^*D^{**}$ with the Dirichlet boundary condition and the Laplacian $-\Delta_N = D^{**}D^*$ with the Neumann boundary condition. It is possible to show that these are self-adjoint and that $(I - \Delta_D)^{-1}$ and $(I - \Delta_N)^{-1}$ are compact operators by computing eigenvalues and eigenfunctions concretely. However, since these facts hold for the Laplacian of a general bounded domain of $\mathbb{R}^n$ (with a nice property), here we give a proof without relying on concrete computations to provide the reader with a flavor of the proof of the general case. For this purpose, we first make some preparations.

Lemma 10.4. *Assume that $A \in \mathbf{B}(\mathcal{H})_{\text{sa}}$ satisfies $\ker A = \{0\}$. Then, the operator T on $\mathcal{H}$ given by $\mathcal{D}(T) = \mathcal{R}(A)$ and $Ax \mapsto x$ is self-adjoint.*

Proof. Since $\mathcal{R}(A)^{\perp} = \ker A = \{0\}$, the domain $\mathcal{D}(T)$ is dense in $\mathcal{H}$. We define $\mathcal{W} \in \mathcal{U}(\mathcal{H} \oplus \mathcal{H})$ by $\mathcal{W}(x, y) = (y, x)$. Then,

$$\mathcal{G}(T) = \{(x, Tx) \in \mathcal{H} \oplus \mathcal{H}; \ x \in \mathcal{D}(T)\} = \{(Ay, y) \in \mathcal{H} \oplus \mathcal{H}; \ y \in \mathcal{H}\}$$
$$= \mathcal{W}\mathcal{G}(A).$$

Thus, we get

$$\mathcal{G}(T^*) = \mathcal{V}\mathcal{G}(T)^{\perp} = \mathcal{V}\mathcal{W}\mathcal{G}(A)^{\perp} = \mathcal{V}\mathcal{W}\mathcal{V}\mathcal{G}(A^*) = \mathcal{V}\mathcal{W}\mathcal{V}\mathcal{W}\mathcal{G}(T)$$
$$= \mathcal{G}(T).$$

Here, we used $\mathcal{V}\mathcal{W}\mathcal{V}\mathcal{W} = I$. $\square$

Theorem 10.3. *Let T be a closed operator from $\mathcal{H}_1$ to $\mathcal{H}_2$ with a dense domain. Let J_T be the embedding from $(\mathcal{D}(T), \langle \cdot, \cdot \rangle_T)$ into $\mathcal{H}_1$. Then, $I + T^*T$ is an invertible self-adjoint operator, and $(I + T^*T)^{-1} = J_T J_T^*$ holds. If J_T is a compact operator, then $(I + T^*T)^{-1}$ is a compact self-adjoint operator.*

Proof. We first show that $I + T^*T$ is a bijection from $\mathcal{D}(T^*T)$ onto $\mathcal{H}_1$. For $x \in \mathcal{D}(T^*T)$, we have $\langle (I + T^*T)x, x \rangle = \|x\|^2 + \|Tx\|^2$, and $I + T^*T$ is an injection. Let $z \in \mathcal{H}_1$. Since

$$\mathcal{H}_1 \oplus \mathcal{H}_2 = \mathcal{G}(T) \oplus \mathcal{G}(T)^{\perp} = \mathcal{G}(T) \oplus \mathcal{V}^{-1}\mathcal{G}(T^*),$$

there exist $x \in \mathcal{D}(T)$ and $y \in \mathcal{D}(T^*)$ satisfying

$$(z, 0) = (x, Tx) + \mathcal{V}^{-1}(y, T^*y) = (x, Tx) + (T^*y, -y).$$

Thus, $z = x + T^*y$ and $0 = Tx - y$, which imply $x \in \mathcal{D}(T^*T)$ and $z = x + T^*Tx$. This shows that $I + T^*T$ is a surjection.

Next, we compute J_T^*. Since $J_T \in \mathbf{B}((\mathcal{D}(T), \langle \cdot, \cdot \rangle_T), \mathcal{H}_1)$, we have $J_T^* \in \mathbf{B}(\mathcal{H}_1, (\mathcal{D}(T), \langle \cdot, \cdot \rangle_T))$. For $y \in \mathcal{H}_1$, let $J_T^*y = z \in \mathcal{D}(T)$. Then, for every $x \in \mathcal{D}(T)$, we have

$$\langle x, y \rangle = \langle J_T x, y \rangle = \langle x, J_T^*y \rangle_T = \langle x, z \rangle_T = \langle x, z \rangle + \langle Tx, Tz \rangle.$$

Since $\langle x, y - z \rangle = \langle Tx, Tz \rangle$ holds for every $x \in \mathcal{D}(T)$, we get $z \in \mathcal{D}(T^*T)$ and $y - z = T^*Tz$. Thus, $J_T J_T^*y = z = (I + T^*T)^{-1}y$ holds. Since $J_T J_T^* \in \mathbf{B}(\mathcal{H}_1)_{\text{sa}}$, the previous lemma shows that $I + T^*T$ is self-adjoint. $\square$

Definition 10.6. We define the *Sobolev space* $W^{n,2}(0,1)$, $n \in \mathbb{N}_0$, as follows. For $n = 0$, we let $W^{0,2}(0,1) = L^2(0,1)$. For $n \in \mathbb{N}$, we let

$$W^{n,2}(0,1) = \{f \in C^{n-1}[0,1];\ f^{(n-1)} \in \mathrm{AC}[0,1],\ f^{(n)} \in L^2(0,1)\},$$

and we define its inner product by $\langle f, g \rangle_{W^{n,2}} = \sum_{k=0}^{n} \langle f^{(k)}, g^{(k)} \rangle$. The space $W^{1,2}(0,1)$ is nothing but $\mathcal{D}(D^*)$ regarded as a Hilbert space with the graph inner product.

Lemma 10.5. *The embedding from $W^{1,2}(0,1)$ into $L^2(0,1)$ is a compact operator.*

Proof. Let $f \in B_{W^{1,2}(0,1)}$. For $0 \le s < t \le 1$, we have

$$|f(t) - f(s)| = \left| \int_s^t f'(r)dr \right| \le \sqrt{t-s}\|f'\|_2 \le \sqrt{t-s},$$

and $B_{W^{1,2}(0,1)}$ is equicontinuous. Also, we have $\|f\|_\infty \le 2$. Indeed, if there existed a point s with $|f(s)| > 2$, the above estimate shows $|f(t)| > 1$ for every $t \in [0,1]$, which contradicts $\|f\|_2 \le 1$. Thus, the Ascoli–Arzelà theorem implies that $B_{W^{1,2}(0,1)}$ is relatively compact in $C[0,1]$. Since the embedding from $C[0,1]$ into $L^2(0,1)$ is continuous, we conclude that $B_{W^{1,2}(0,1)}$ is relatively compact in $L^2(0,1)$. $\square$

Example 10.4. We define a symmetric operator L on $L^2(0,1)$ by $\mathcal{D}(L) = C_c^\infty(0,1)$, $Lf = -f''$. Since $T = D^{**}, D^*, D_c$, $c \in \mathbb{C} \cup \{\infty\}$ in Problem 10.1 are all closed operators, the operators T^*T are self-adjoint extensions of L. Also, since the embedding from $(\mathcal{D}(T), \langle \cdot, \cdot \rangle_T)$ into $W^{1,2}(0,1)$ is an isometry and that from $W^{1,2}(0,1)$ into $L^2(0,1)$ is a compact operator, we see that $(I + T^*T)^{-1}$ is a compact operator. Therefore, $\sigma(T^*T) = \sigma_p(T^*T)$ holds, and T^*T has a CONS consisting of its eigenfunctions.

Problem 10.2. Determine $\sigma(T^*T)$ for $T = D^{**}, D^*, D_c$, $c \in \mathbb{C} \cup \{\infty\}$.

We leave, as Exercise 10.2, the problem of determining L^* and all the self-adjoint extensions of L by using Lemma 10.3 as in Theorem 10.2.

10.3 Cayley Transform of Symmetric Operators

We assume that S is a symmetric operator on $\mathcal{H}$ throughout this section. Since $\langle Sx, x \rangle = \langle x, Sx \rangle$ holds for every $x \in \mathcal{D}(S)$, we have $\langle Sx, x \rangle \in \mathbb{R}$ and, in particular, $\sigma_{\mathrm{p}}(S) \subset \mathbb{R}$. Thus, $S \pm iI : \mathcal{D}(S) \to \mathcal{H}$ is injective. Also,

$$\|(S \pm iI)x\|^2 = \|x\|^2 + \|Sx\|^2 + 2\,\mathrm{Re}\langle Sx, \pm ix \rangle = \|x\|^2 + \|Sx\|^2$$

shows that $\|(S+iI)x\| = \|(S-iI)x\|$ holds.

Definition 10.7. We define the *Cayley transform* V_S of a symmetric operator S on $\mathcal{H}$ by $\mathcal{D}(V_S) = \mathcal{R}(S+iI)$, $(S+iI)x \mapsto (S-iI)x$. The Cayley transform V_S is an isometry from $\mathcal{D}(V_S) = \mathcal{R}(S+iI)$ onto $\mathcal{R}(V_S) = \mathcal{R}(S-iI)$.

For $x \in \mathcal{D}(S)$, let $y = (S+iI)x$. Then, $V_S y = (S-iI)x$ implies $x = 2i^{-1}(y - V_S y)$ and $Sx = 2^{-1}(y + V_S y)$. Thus, $\mathcal{R}(I - V_S)$ is dense in $\mathcal{H}$.

Theorem 10.4. *The Cayley transform is a bijection between the following two sets of operators preserving extension:*

(1) *the set of symmetric operators on $\mathcal{H}$;*
(2) *the set of operators V on $\mathcal{H}$ such that $V : \mathcal{D}(V) \to \mathcal{H}$ is an isometry and $\overline{\mathcal{R}(I - V)} = \mathcal{H}$.*

Proof. Assume that V is an operator satisfying the condition in (2). Note that $\ker(I - V) = \{0\}$ holds. Indeed, if $x \in \mathcal{D}(V)$ satisfies $Vx = x$, for every $y \in \mathcal{D}(V)$, we have

$$\langle (I-V)y, x \rangle = \langle y, x \rangle - \langle Vy, x \rangle = \langle y, x \rangle - \langle Vy, Vx \rangle = 0,$$

which shows $x = 0$. Using this fact, we can define an operator S_V by $\mathcal{D}(S_V) = \mathcal{R}(I - V)$ and $(I - V)y \mapsto i(I + V)y$. Then, $\mathcal{D}(S_V)$ is dense in $\mathcal{H}$. For $y_1, y_2 \in \mathcal{D}(V)$, we have

$$\langle S_V(I-V)y_1, (I-V)y_2 \rangle = \langle i(I+V)y_1, (I-V)y_2 \rangle$$
$$= i(\langle Vy_1, y_2 \rangle - \langle y_1, Vy_2 \rangle),$$
$$\langle (I-V)y_1, S_V(I-V)y_2 \rangle = \langle (I-V)y_1, i(I+V)y_2 \rangle$$
$$= -i(-\langle Vy_1, y_2 \rangle + \langle y_1, Vy_2 \rangle),$$

and S_V is a symmetric operator. We have already seen that $S_{V_S} = S$. A similar argument shows $V_{S_V} = V$. $\qquad\square$

Lemma 10.6. *For a symmetric operator* S *on* $\mathcal{H}$, *the equality* $\overline{\mathcal{R}(S \pm iI)} = \mathcal{R}(\overline{S} \pm iI)$ *holds. In particular,*

$$S \text{ is closed } \iff \mathcal{R}(S + iI) \text{ is closed } \iff \mathcal{R}(S - iI) \text{ is closed.}$$

Proof. We show only $\overline{\mathcal{R}(S + iI)} = \mathcal{R}(\overline{S} + iI)$. For $y \in \overline{\mathcal{R}(S + iI)}$, there exists a sequence $\{x_n\}_{n=1}^\infty$ in $\mathcal{D}(S)$ such that $\{(S + iI)x_n\}_{n=1}^\infty$ converges to y. Since

$$\|(S + iI)x_n - (S + iI)x_m\|^2 = \|x_n - x_m\|^2 + \|Sx_n - Sx_m\|^2,$$

$\{x_n\}_{n=1}^\infty$ and $\{Sx_n\}_{n=1}^\infty$ are Cauchy sequences and converge. Since S is closable, we have $x = \lim_{n\to\infty} x_n \in \mathcal{D}(\overline{S})$ and $\lim_{n\to\infty} Sx_n = \overline{S}x$. Thus, $y \in \mathcal{R}(\overline{S} + iI)$.

For $x \in \mathcal{D}(\overline{S})$, there exists a sequence $\{x_n\}_{n=1}^\infty$ in $\mathcal{D}(S)$ such that $\{x_n\}_{n=1}^\infty$ converges to x and $\{Sx_n\}_{n=1}^\infty$ converges to $\overline{S}x$. Thus, $(\overline{S} + iI)x \in \overline{\mathcal{R}(S + iI)}$. $\qquad\square$

Lemma 10.7. *For a symmetric operator* S *on* $\mathcal{H}$,

$$\mathcal{D}(S^*) = \mathcal{D}(\overline{S}) \oplus \ker(S^* - iI) \oplus \ker(S^* + iI)$$

holds in the sense of an orthogonal direct sum with respect to the graph inner product of S^*. *In particular,* S *is essentially self-adjoint if and only if* $\ker(S^* - iI) = \ker(S^* + iI) = \{0\}$.

Proof. Throughout this proof, orthogonality is with respect to the graph inner product of S^*. Since $\mathcal{D}(\overline{S})$, $\ker(S^* - iI)$, and $\ker(S^* + iI)$ are mutually orthogonal, it suffices to show

$$\mathcal{D}(S^*) \cap \mathcal{D}(S)^\perp \subset \ker(S^* - iI) + \ker(S^* + iI).$$

Let $y \in \mathcal{D}(S^*) \cap \mathcal{D}(S)^\perp$. Then, for every $x \in \mathcal{D}(S)$, we have

$$0 = \langle x, y \rangle + \langle S^*x, S^*y \rangle = \langle x, y \rangle + \langle Sx, S^*y \rangle.$$

Thus, $S^*y \in \mathcal{D}(S^*)$ and $S^{*2}y = -y$ holds. The lemma follows from

$$y = \frac{1}{2}(y - iS^*y) + \frac{1}{2}(y + iS^*y) \in \ker(S^* - iI) + \ker(S^* + iI).$$

$\square$

The following theorem plays an essential role in the proof of the spectral decomposition theorem for unbounded self-adjoint operators shown in the following section.

Theorem 10.5. *For a symmetric operator S on $\mathcal{H}$, the following conditions are equivalent:*

(1) *S is self-adjoint.*
(2) *V_S is a unitary operator, that is, $\mathcal{R}(S \pm iI) = \mathcal{H}$.*

Proof. (1) $\implies$ (2). Since S is a closed operator, the ranges $\mathcal{R}(S \pm iI)$ are closed. Since $\mathcal{R}(S \pm iI)^{\perp} = \ker(S \mp iI) = \{0\}$, we get $\mathcal{R}(S \pm iI) = \mathcal{H}$.

(2) $\implies$ (1). Since $\mathcal{R}(S \pm iI) = \mathcal{H}$ is closed, we have $S = \overline{S}$, and $\{0\} = \mathcal{R}(S \pm iI)^{\perp} = \ker(S^* \mp iI)$. Thus, the previous lemma shows $\mathcal{D}(S^*) = \mathcal{D}(S)$. $\square$

Corollary 10.2. *If A is self-adjoint, we have $\sigma(A) \subset \mathbb{R}$.*

Proof.   Let $\lambda = \xi + i\eta$, $\xi, \eta \in \mathbb{R}$, $\eta \neq 0$, and let $B = \eta^{-1}(\xi I - A)$. Then, B is self-adjoint as well, and $\lambda I - A = \eta(B + iI)$. Since $B + iI$ is a closed operator and a bijection from $\mathcal{D}(A)$ onto $\mathcal{H}$, it is invertible. $\square$

We summarize what we have shown so far as follows.

Theorem 10.6. *Let S be a symmetric operator on $\mathcal{H}$. Then, the Cayley transform gives a one-to-one correspondence between the following two sets of operators:*

• *the set of symmetric extensions of S;*
• *the set of isometric extensions of V_S.*

Moreover, if S is closed, then there exists a one-to-one correspondence between the latter and the following set:

• *the set of pairs (K, W) of a closed subspace K of $\ker(S^* - iI)$ and an isometry $W : K \to \ker(S^* + iI)$.*

In particular, the restriction of the Cayley transform to $\ker(S^* - iI)$ gives a one-to-one correspondence between the set of self-adjoint extensions of S and unitary operators from $\ker(S^* - iI)$ onto $\ker(S^* + iI)$.

Corollary 10.3. *For a symmetric operator S, the following conditions are equivalent:*

(1) *There exists a self-adjoint extension of S.*
(2) $\dim(\ker(S^* - iI)) = \dim(\ker(S^* + iI))$.

We call $(\dim \ker(S^* - iI), \dim \ker(S^* + iI)) \in (\mathbb{N}_0 \cup \{\infty\})^2$ the *deficiency indices* or *defect numbers* of S.

Example 10.5. The deficiency indices of D in Example 10.2 are $(1,1)$. Let $\varphi_+(t) = e^{1-t}$, and let $\varphi_-(t) = e^t$. Then, $\ker(D^* - iI) = \mathbb{C}\varphi_+$ and $\ker(D^* + iI) = \mathbb{C}\varphi_-$. Since $\|\varphi_+\|_2 = \|\varphi_-\|_2$, an isometric extension V of $V_{D^{**}}$ is characterized by $V\varphi_+ = \zeta\varphi_-$, with $|\zeta| = 1$. Let D_c be the self-adjoint extension corresponding to V. Then, since $(I - V)\varphi_+ = \varphi_+ - \zeta\varphi_- \in \mathcal{D}(D_c)$, the number c is determined by

$$c = \frac{\varphi_+(1) - \zeta\varphi_-(1)}{\varphi_+(0) - \zeta\varphi_-(0)} = \frac{1 - \zeta e}{e - \zeta}.$$

Example 10.6. The deficiency indices of $\tilde{D}$ in Example 10.3 are $(1,0)$, and $\ker(\tilde{D}^* - iI) = \mathbb{C}e^{-t}$. This shows again that $\tilde{D}$ has no self-adjoint extension.

10.4 Spectral Decomposition Theorem of Unbounded Self-Adjoint Operators

Let $E = \{E_t\}_{t\in\mathbb{R}}$ be a spectral family consisting of projections on $\mathcal{H}$. We use the notation in Section 9.2. Let Ω be the set of increasing points of $\{E_t\}_{t\in\mathbb{R}}$. Then, there exists an algebra homomorphism π_E from the set of bounded Borel functions $\mathcal{B}^b(\Omega)$ on Ω to $\mathbf{B}(\mathcal{H})$ of algebras over $\mathbb{C}$. Let $\mathcal{B}(\Omega)$ be the set of Borel functions on Ω. In order to formulate the spectral decomposition of an unbounded self-adjoint operator, we first define $\pi_E(f) = \int_\Omega f(t)dE_t$ for $f \in \mathcal{B}(\Omega)$ as a closed operator.

We fix $f \in \mathcal{B}(\Omega)$, and for $n \in \mathbb{N}$, we set $\Omega_n = \{\omega \in \Omega; |f(\omega)| \leq n\}$. Then, $\bigcup_{n=1}^\infty \Omega_n = \Omega$ holds. Let $f_n = \chi_{\Omega_n} f$.

Lemma 10.8. *Under the assumption as above, the following hold:*

(1) s-$\lim_{n\to\infty} \pi_E(\chi_{\Omega_n}) = I$.

(2) *Let* $\mathcal{K} = \{x \in \mathcal{H}; \int_\Omega |f(t)|^2 d\langle E_t x, x\rangle < \infty\}$. *Then,* $\mathcal{K}$ *is a dense subspace of* $\mathcal{H}$.

(3) *For every* $x \in \mathcal{K}$, *the sequence* $\{\pi_E(f_n)x\}_{n=1}^\infty$ *converges.*

Proof. (1) follows from

$$\|x - \pi_E(\chi_{\Omega_n})x\|^2 = \int_\Omega |1 - \chi_{\Omega_n}(t)|^2 d\mu_x(t) = \mu_x(\Omega \backslash \Omega_n) \to 0, \quad (n \to \infty).$$

(2) For every $n \in \mathbb{N}$, we have $\mathcal{R}(\pi_E(\chi_{\Omega_n})) \subset \mathcal{K}$, and $\mathcal{K}$ is dense in $\mathcal{H}$. Since we can also see $\mathbb{C}\mathcal{K} \subset \mathcal{K}$ from the definition, it suffices to show that $x, y \in \mathcal{K}$ implies $x + y \in \mathcal{K}$. From

$$\langle E_t(x+y), x+y\rangle + \langle E_t(x-y), x-y\rangle = 2\langle E_t x, x\rangle + 2\langle E_t y, y\rangle,$$

we have $\mu_{x+y} + \mu_{x-y} = 2\mu_x + 2\mu_y$, and

$$\int_\Omega |f(t)|^2 d\mu_{x+y}(t) \le 2 \int_\Omega |f(t)|^2 d\mu_x(t) + 2 \int_\Omega |f(t)|^2 d\mu_y(t) < \infty,$$

which shows $x + y \in \mathcal{K}$.

(3) For $x \in \mathcal{K}$ and $m > n$, we have

$$\|\pi_E(f_m)x - \pi_E(f_n)x\|^2 = \int_\Omega |\chi_{\Omega_m} - \chi_{\Omega_n}|^2 |f(t)|^2 d\mu_x(t)$$

$$= \int_{\Omega_m} |f(t)|^2 d\mu_x(t) - \int_{\Omega_n} |f(t)|^2 d\mu_x(t),$$

which shows that $\{\pi_E(f_n)x\}_{n=1}^\infty$ is a Cauchy sequence and it converges. $\square$

Definition 10.8. For $f \in \mathcal{B}(\Omega)$, we define an operator $\pi_E(f)$ on $\mathcal{H}$ by

$$\mathcal{D}(\pi_E(f)) = \{x \in \mathcal{H}; \int_\Omega |f(t)|^2 d\langle E_t x, x\rangle < \infty\},$$

and $\pi_E(f)x = \lim_{n\to\infty} \pi_E(f_n)x$. We sometimes denote $\pi_E(f)$ by $\int_\Omega f(t)dE_t$ or $\int_\mathbb{R} f(t)dE_t$.

Theorem 10.7. *For $f \in \mathcal{B}(\Omega)$, the following hold:*

(1) $\pi_E(f)$ *is a closed operator, and $\pi_E(f)^* = \pi_E(\overline{f})$ holds. In particular, $\pi_E(f)$ is self-adjoint if f is real-valued.*

(2) *If $T \in \mathbf{B}(\mathcal{H})$ commutes with E_t for every $t \in \mathbb{R}$, we have $T\pi_E(f) \subset \pi_E(f)T$.*

Proof. (1) We first show $\pi_E(\overline{f}) \subset \pi_E(f)^*$. Note that we have $\mathcal{D}(\pi_E(f)) = \mathcal{D}(\pi_E(\overline{f}))$. For $x, y \in \mathcal{D}(\pi_E(f))$, we have

$$\langle \pi_E(f)x, y \rangle = \lim_{n \to \infty} \langle \pi_E(f_n)x, y \rangle = \lim_{n \to \infty} \langle x, \pi_E(\overline{f_n})y \rangle = \langle x, \pi_E(\overline{f})y \rangle,$$

and $\pi_E(\overline{f}) \subset \pi_E(f)^*$ holds.

Next, let $y \in \mathcal{D}(\pi_E(f)^*)$ and $y^* = \pi_E(f)^*y$. Let $P_n = \pi_E(\chi_{\Omega_n})$. Note that $P_n\mathcal{D}(\pi_E(f)) \subset \mathcal{D}(\pi_E(f))$ holds. For $x \in \mathcal{D}(\pi_E(f))$, we have $\langle \pi_E(f)P_nx, y \rangle = \langle x, P_ny^* \rangle$. On the other hand,

$$\langle \pi_E(f)P_nx, y \rangle = \langle \pi_E(f_n)x, y \rangle = \langle x, \pi_E(\overline{f_n})y \rangle.$$

Since $\mathcal{D}(\pi_E(f))$ is dense in $\mathcal{H}$, we get $P_ny^* = \pi_E(\overline{f_n})y$, and

$$\|y^*\|^2 = \lim_{n \to \infty} \|P_ny^*\|^2 = \lim_{n \to \infty} \|\pi_E(\overline{f_n})y\|^2 = \lim_{n \to \infty} \int_{\Omega_n} |f(t)|^2 d\mu_y(t).$$

By the monotone convergence theorem, we get $\int_{\Omega} |f(t)|^2 d\mu_y(t) = \|y^*\|^2 < \infty$, and $y \in \mathcal{D}(\pi_E(\overline{f}))$.

(2) We first see that $\mu_{Tx} \leq \|T\|^2 \mu_x$ holds for every $x \in \mathcal{H}$. Indeed, let $F \subset \Omega$ be a Borel set, and let $P = \pi_E(\chi_F)$. Then, we have

$$\mu_{Tx}(F) = \|PTx\|^2 = \|TPx\|^2 \leq \|T\|^2\|Px\|^2 = \|T\|^2 \mu_x(F).$$

For $x \in \mathcal{D}(\pi_E(f))$, we have

$$\int_{\Omega} |f(t)|^2 d\mu_{Tx}(t) \leq \|T\|^2 \int_{\Omega} |f(t)|^2 d\mu_x(t) < \infty,$$

and $Tx \in \mathcal{D}(\pi_E(f))$. Thus, we get

$$\pi_E(f)Tx = \lim_{n \to \infty} \pi_E(f_n)Tx = \lim_{n \to \infty} T\pi_E(f_n)x = T\pi_E(f)x. \qquad \square$$

Lemma 10.9. *For every* $x \in \mathcal{D}(\pi_E(f))$, *we have* $\langle \pi_E(f)x, x \rangle = \int_\Omega f(t) d \langle E_t x, x \rangle$.

Proof. Since μ_x is a finite measure and $f \in L^2(\Omega, \mu_x)$, we have $f \in L^1(\Omega, \mu_x)$. Thus,

$$\langle \pi_E(f)x, x \rangle = \lim_{n \to \infty} \langle \pi_E(f_n)x, x \rangle = \lim_{n \to \infty} \int_\Omega f_n(t) d\mu_x(t)$$

$$= \int_\Omega f(t) d\mu_x(t)$$

holds. $\qquad \square$

Lemma 10.10. *For* $f, g \in \mathcal{B}(\Omega)$, *the following hold*:

(1) $\pi_E(f)\pi_E(g) \subset \pi_E(fg)$.
(2) $\pi_E(\bar{f})\pi_E(f) = \pi_E(|f|^2)$.

Proof. (1) Let $x \in \mathcal{D}(\pi_E(f)\pi_E(g))$. Then, $x \in \mathcal{D}(\pi_E(g))$ and $y = \pi_E(g)x \in \mathcal{D}(\pi_E(f))$. Since

$$\langle E_t y, y \rangle = \langle E_t \pi_E(g)x, \pi_E(g)x \rangle = \|\pi_E(g)E_t x\|^2 = \lim_{n \to \infty} \|\pi_E(g_n)E_t x\|^2$$

$$= \lim_{n \to \infty} \int_{(-\infty, t] \cap \Omega} |g_n(s)|^2 d\mu_x(s) = \int_{(-\infty, t] \cap \Omega} |g(s)|^2 d\mu_x(s),$$

we get

$$\int_\Omega |f(t)|^2 d\mu_y(t) = \int_\Omega |f(t)g(t)|^2 d\mu_x(t),$$

and $x \in \pi_E(fg)$. Let $\Omega'_n = \{ t \in \Omega; \ |g(t)| \leq n \}$, and let $Q_n = \pi_E(\chi_{\Omega_n \cap \Omega'_n})$. Then, $\bigcup_{n=1}^\infty \mathcal{R}(Q_n)$ is dense in $\mathcal{H}$. For $z \in \mathcal{R}(Q_n)$, we have

$$\langle \pi_E(f)\pi_E(g)x, z \rangle = \langle Q_n \pi_E(f)\pi_E(g)x, z \rangle = \langle \pi_E(f_n)Q_n \pi_E(g)x, z \rangle$$

$$= \langle \pi_E(f_n)\pi_E(g_n)x, z \rangle = \langle \pi_E(f_n g_n)x, z \rangle.$$

On the other hand, we have

$$\langle \pi_E(fg)x, z \rangle = \langle Q_n \pi_E(fg)x, z \rangle = \langle \pi_E(f_n g_n)x, z \rangle,$$

and $\pi_E(f)\pi_E(g)x = \pi_E(fg)x$ holds.

(2) The above argument shows $\pi_E(f)^*\pi_E(f) \subset \pi_E(|f|^2)$, and Theorems 10.3 and 10.7 show that both sides are self-adjoint. Taking the adjoint, we get $\pi_E(|f|^2) \subset \pi_E(f)^*\pi_E(f)$. $\qquad\qquad\Box$

Let A be a self-adjoint operator on $\mathcal{H}$, and let U be the Cayley transform $(A-iI)(A+iI)^{-1}$ of A. Then, $U \in \mathcal{U}(\mathcal{H})$ and $\ker(I-U) = \{0\}$. Let $U = \int_{\mathbb{R}} e^{is}dE_s^U$ be the spectral decomposition of U. Since $E_0^U = 0$ and $E_{2\pi-0}^U = I$, we write $U = \int_{(0,2\pi)} e^{is}dE_s^U$. We consider the homeomorphism $\mathbb{R} \to (0,2\pi)$, $t \mapsto s$, $(t-i)/(t+i) = e^{is}$, and $t = -\cot(s/2)$. Since $s \to +0$ as $t \to -\infty$, if we set $E_t^A = E_s^U$, we get a spectral family: $E^A = \{E_t^A\}_{t\in\mathbb{R}}$.

Theorem 10.8 (Spectral decomposition). *For a self-adjoint operator A on $\mathcal{H}$, we have $A = \int_{\mathbb{R}} tdE_t^A$.*

Proof. For simplicity, we write $E_t^A = E_t$ and $E_s^U = F_s$. Let $B = \int_{\mathbb{R}} tdE_t$. Then, B is a self-adjoint operator. We show $A \subset B$ as follows. Taking the adjoint of both sides, we get $B = B^* \subset A^* = A$, and $A = B$.

For $x \in \mathcal{D}(A)$, let $y = (A+iI)x$. Then, $x = 2i^{-1}(I-U)y$ and $Ax = 2^{-1}(I+U)y$. We show $\mathcal{D}(A) \subset \mathcal{D}(B)$. Indeed, since

$$\int_{\mathbb{R}} t^2 d\langle E_t x, x\rangle = \int_{(0,2\pi)} \left| i\frac{1+e^{is}}{1-e^{is}} \right|^2 d\left\langle F_s \frac{1}{2i}(I-U)y, \frac{1}{2i}(I-U)y \right\rangle$$

$$= \frac{1}{4}\int_{(0,2\pi)} |1+e^{is}|^2 d\langle F_s y, y\rangle < \infty,$$

we obtain $x \in \mathcal{D}(B)$. We have

$$Bx = \lim_{n\to\infty} \pi_E(\chi_{[-n,n]}(t)t)x$$

$$= \lim_{m\to\infty} \pi_F\left(\chi_{[\frac{1}{m},2\pi-\frac{1}{m}]}(s)i\frac{1+e^{is}}{1-e^{is}}\right)\frac{1}{2i}(I-U)y$$

$$= \lim_{m\to\infty} \pi_F(\chi_{[\frac{1}{m},2\pi-\frac{1}{m}]}(s)(1+e^{is}))\frac{1}{2}y$$

$$= \lim_{m\to\infty} \pi_F(\chi_{[\frac{1}{m},2\pi-\frac{1}{m}]}(s))\frac{1}{2}(I+U)y$$

$$= Ax,$$

which shows $A \subset B$. $\qquad\qquad\Box$

As in the case of bounded operators, the uniqueness of the spectral decomposition holds. Proposition 9.1 holds as well. We denote $\pi_{E^A}(f)$ by $f(A)$.

While we have constructed the spectral family of A by using the Cayley transform of A, there are several other ways to do it. When $\mathcal{H}$ is a real Hilbert space, the above proof using the Cayley transform needs modification because the spectral decomposition of unitary operators cannot be performed within $\mathcal{H}$. One way is to apply the above argument to the complex Hilbert space $\mathcal{H} \otimes \mathbb{C} = \mathcal{H} + i\mathcal{H}$ and show that E_t^A leaves the original $\mathcal{H}$ invariant. Or, alternatively, we can construct the spectral family of A by using that of $A(I+A^2)^{-\frac{1}{2}} \in \mathbf{B}(\mathcal{H}_{\mathrm{sa}})$ (see Schmüdgen, 2012).

10.5 Applications of the Spectral Decomposition Theorem

10.5.1 *Polar decomposition of a closed operator*

If a self-adjoint operator A on a Hilbert space $\mathcal{H}$ satisfies $\langle Ax, x \rangle \geq 0$ for every $x \in \mathcal{D}(A)$, we say that A is a *positive operator*. Since $\langle Ax, x \rangle = \int_{\mathbb{R}} t d\langle E_t^A x, x \rangle$ holds, this condition is equivalent to $\sigma(A) \subset [0, \infty)$, as in the case of bounded operators.

Theorem 10.9. *For every positive self-adjoint operator A on $\mathcal{H}$, there exists a unique positive self-adjoint operator B satisfying $A = B^2$.*

Proof. Let $B = \int_{[0,\infty)} \sqrt{t} dE_t^A$. Then, Lemma 10.10 shows $B^2 = A$. On the other hand, if B satisfies the condition, we have $A = \int_{[0,\infty)} s^2 dE_s^B$, and the uniqueness of the spectral decomposition shows $E_t^A = E_{\sqrt{t}}^B$. Thus, B is uniquely determined by A. $\qquad\square$

We denote the above B by $\sqrt{A}$ or $A^{1/2}$ and call it the *square root operator* of A.

Let T be a closed operator from $\mathcal{H}_1$ to $\mathcal{H}_2$ with dense domain. Then, Theorem 10.3 shows that T^*T is self-adjoint. For $x \in \mathcal{D}(T^*T)$, we have $\langle T^*Tx, x \rangle = \langle Tx, Tx \rangle$, and T^*T is a positive operator. We denote $(T^*T)^{1/2}$ by $|T|$ and call it the *absolute value operator* of T.

If $\mathcal{K}$ is a subspace of $\mathcal{D}(T)$ and the closure of the restriction of T to $\mathcal{K}$ coincides with T, we say that $\mathcal{K}$ is a *core* for T. This is equivalent to the condition that $\mathcal{K}$ is dense in $\mathcal{D}(T)$ with respect to the graph inner product $\langle \cdot, \cdot \rangle_T$.

Lemma 10.11. *If T is a closed operator from $\mathcal{H}_1$ to $\mathcal{H}_2$ with a dense domain, then $\mathcal{D}(T^*T)$ is a core for T.*

Proof. Let $y \in \mathcal{D}(T) \cap \mathcal{D}(T^*T)^{\perp}$ with respect to the graph inner product $\langle \cdot, \cdot \rangle_T$. Then, for every $x \in \mathcal{D}(T^*T)$, we have

$$0 = \langle x, y \rangle_T = \langle x, y \rangle + \langle Tx, Ty \rangle = \langle x, y \rangle + \langle T^*Tx, y \rangle.$$

Since $\mathcal{R}(I + T^*T) = \mathcal{H}_1$, we get $y = 0$. $\qquad\qquad\square$

From the above lemma, we see that $\mathcal{D}(T^*T)$ is a core for $|T|$ as well.

Theorem 10.10. *Let T be a closed operator from $\mathcal{H}_1$ to $\mathcal{H}_2$ with a dense domain. Then, there exists a unique partial isometry, $W \in \mathbf{B}(\mathcal{H}_1, \mathcal{H}_2)$, satisfying $\ker T = \ker W$ and $T = W|T|$. Further, we have $W^*T = |T|$ and $|T^*| = W|T|W^*$.*

Proof. Once we show that $\mathcal{D}(T) = \mathcal{D}(|T|)$ and $\|Tx\| = \||T|x\|$ hold for every $x \in \mathcal{D}(T)$, the rest of the argument is the same as in the case of bounded operators.

First, for $x \in \mathcal{D}(T^*T) = \mathcal{D}(|T|^2)$, we have

$$\|Tx\|^2 = \langle T^*Tx, x \rangle = \langle |T|^2 x, x \rangle = \||T|x\|^2.$$

Note that the graph inner product of T and that of $|T|$ coincide on $\mathcal{D}(T^*T)$. Since $\mathcal{D}(T^*T)$ is a core for T, for every $x \in \mathcal{D}(T)$, there exists a sequence $\{x_n\}_{n=1}^{\infty}$ in $\mathcal{D}(T^*T)$ such that $\{x_n\}_{n=1}^{\infty}$ converges to x and $\{Tx_n\}_{n=1}^{\infty}$ converges to Tx. Since $\{x_n\}$ is a Cauchy sequence with respect to the graph inner product of $|T|$, we have $x \in \mathcal{D}(|T|)$, and $\{|T|x_n\}_{n=1}^{\infty}$ converges to $|T|x$. Since $\|Tx_n\| = \||T|x_n\|$, we get $\|Tx\| = \||T|x\|$, which shows $\mathcal{D}(T) \subset \mathcal{D}(|T|)$. Switching the roles of T and $|T|$, we also get $\mathcal{D}(|T|) \subset \mathcal{D}(T)$. $\qquad\square$

10.5.2　*The Stone theorem*

A *one-parameter unitary group* $U = \{U(t)\}_{t\in\mathbb{R}}$ is a group homomorphism $U : \mathbb{R} \to \mathcal{U}(\mathcal{H})$ from the additive group of all real numbers $\mathbb{R}$ to the unitary group $\mathcal{U}(\mathcal{H})$ of a Hilbert space $\mathcal{H}$. If, moreover, $t \mapsto U(t)$ is continuous with respect to the strong operator topology (that is, the map $t \mapsto U(t)x$ is continuous for every $x \in \mathcal{H}$), we say that U is strongly continuous. If A is a self-adjoint operator on $\mathcal{H}$ and $U(t) = e^{itA} = \int_{\mathbb{R}} e^{it\lambda} dE_\lambda^A$, then U is a one-parameter unitary group, and

$$\|U(s)x - U(t)x\|^2 = \int_{\mathbb{R}} |e^{is\lambda} - e^{it\lambda}|^2 d\langle E_\lambda^A x, x\rangle$$

shows that it is strongly continuous. The Stone theorem shows that its converse holds.

Problem 10.3. Let A be a self-adjoint operator. Show that for every $x \in \mathcal{D}(A)$,

$$\lim_{t\to 0} \frac{1}{it}(e^{itA}x - x) = Ax$$

holds.

Lemma 10.12. *For a self-adjoint operator A on a Hilbert space $\mathcal{H}$, the following holds:*

$$\mathcal{D}(A) = \left\{ x \in \mathcal{H}; \ \exists \lim_{t\to 0} \frac{1}{it}(e^{itA}x - x) \right\}.$$

Proof.　We define an operator T on $\mathcal{H}$ by setting $\mathcal{D}(T)$ to be the set on the right-hand side, and

$$Tx = \lim_{t\to 0} \frac{1}{it}(e^{itA}x - x).$$

For $x, y \in \mathcal{D}(T)$, we have

$$\left\langle \frac{1}{it}(e^{itA}x - x), y \right\rangle = \left\langle x, \frac{1}{-it}(e^{-itA}y - y) \right\rangle,$$

and $\langle Tx, y\rangle = \langle x, Ty\rangle$ holds, which shows that T is a symmetric operator. Since A is self-adjoint and $A \subset T$, we get $A = T$. □

Lemma 10.13. *Let U be a strongly continuous one-parameter unitary group on $\mathcal{H}$.*

(1) *For every $f \in L^1(\mathbb{R})$, there exists a unique $U_f \in \mathbf{B}(\mathcal{H})$ satisfying $\langle U_f x, y \rangle = \int_{\mathbb{R}} f(t) \langle U(t) x, y \rangle dt$ for all $x, y \in \mathcal{H}$. Moreover, $\|U_f\| \leq \|f\|_1$ holds.*

(2) *For $f, g \in L^1(\mathbb{R})$, we have $U_f U_g = U_{f*g}$ and $U_f{}^* = U_{\check{f}}$. Here, $f * g(t) = \int_{\mathbb{R}} f(s) g(t - s) ds$ and $\check{f}(t) = \overline{f}(-t)$. Furthermore, $U(s) U_f = U_{f^s}$ holds, where $f^s(t) = f(t - s)$.*

(3) *The subspace $\mathrm{span}\{U_f x \in \mathcal{H}; \ x \in \mathcal{H}, \ f \in C_c^\infty(\mathbb{R})\}$ is dense in $\mathcal{H}$.*

(4) *For $f \in C_c^\infty(\mathbb{R})$ and $x \in \mathcal{H}$,*

$$\lim_{t \to 0} \frac{1}{it} (U(t) U_f x - U_f x) = i U_{f'} x$$

holds.

Proof. (1) We define a sesquilinear form on $\mathcal{H}$ by $b_f(x, y) = \int_{\mathbb{R}} f(t) \langle U(t) x, y \rangle dt$. Since $|b_f(x, y)| \leq \|f\|_1 \|x\| \|y\|$, the statement follows.

(2) First, since

$$\langle U(s) U_g x, y \rangle = \langle U_g x, U(-s) y \rangle = \int_{\mathbb{R}} g(t) \langle U(t) x, U(-s) y \rangle dt$$

$$= \int_{\mathbb{R}} g(t) \langle U(s + t) x, y \rangle dt = \int_{\mathbb{R}} g^s(t) \langle U(t) x, y \rangle dt,$$

we get $U(s) U_g = U_{g^s}$. From this, we get

$$\langle U_f U_g x, y \rangle = \int_{\mathbb{R}} f(s) \langle U(s) U_g x, y \rangle ds$$

$$= \int_{\mathbb{R}} f(s) \int_{\mathbb{R}} g(t - s) \langle U(t) x, y \rangle dt ds,$$

and the Fubini theorem implies $U_f U_g = U_{f*g}$.

$$\langle U_f{}^* x, y \rangle = \langle x, U_f y \rangle = \overline{\int_{\mathbb{R}} f(t) \langle U(t) y, x \rangle dt} = \int_{\mathbb{R}} \overline{f(t)} \langle U(-t) x, y \rangle dt$$

shows $U_f{}^* = U_{\check{f}}$.

(3) We take $f \in C_c^\infty(\mathbb{R})$ satisfying $\operatorname{supp} f \subset [-1, 1]$, $f(t) \geq 0$, $f(t) = f(-t)$, and $\int_{\mathbb{R}} f(t)dt = 1$. For $\varepsilon > 0$, let $f_\varepsilon(t) = \varepsilon^{-1}f(\varepsilon^{-1}t)$. Then, we have

$$f_\varepsilon = \check{f}_\varepsilon, \quad \|f_\varepsilon\|_1 = \|f_\varepsilon * f_\varepsilon\|_1 = 1,$$

and

$$\operatorname{supp} f_\varepsilon \subset [-\varepsilon, \varepsilon], \quad \operatorname{supp} f_\varepsilon * f_\varepsilon \subset [-2\varepsilon, 2\varepsilon].$$

Since

$$\|U_{f_\varepsilon}x - x\|^2 = \langle U_{f_\varepsilon * f_\varepsilon}x, x\rangle - 2\langle U_{f_\varepsilon}x, x\rangle + \|x\|^2$$

$$= \int_{\mathbb{R}} f_\varepsilon * f_\varepsilon(t)(\langle U(t)x, x\rangle - \|x\|^2)dt$$

$$+ 2\int_{\mathbb{R}} f_\varepsilon(t)(\|x\|^2 - \langle U(t)x, x\rangle)dt$$

$$\leq (\|f_\varepsilon * f_\varepsilon\|_1 + 2\|f_\varepsilon\|_1) \sup_{t\in[-2\varepsilon, 2\varepsilon]} |\langle U(t)x, x\rangle - \|x\|^2|$$

$$= 3 \sup_{t\in[-2\varepsilon, 2\varepsilon]} |\langle U(t)x, x\rangle - \|x\|^2|,$$

we get $\lim_{\varepsilon \to +0} \|U_{f_\varepsilon}x - x\| = 0$.

(4) For $f \in C_c^\infty(\mathbb{R})$, we have

$$\left\|\frac{1}{it}(U(t)U_f x - U_f x) - iU_{f'}x\right\| = \left\|U_{\frac{f^t - f}{t} + f'}x\right\| \leq \left\|\frac{f^t - f}{t} + f'\right\|_1 \|x\|.$$

Since

$$\frac{f^t(s) - f(s)}{t} + f'(s) = \int_0^1 (f'(s) - f'(s - rt))dr$$

$$= t\int_0^1 \int_0^1 rf''(s - prt)dpdr,$$

we get

$$\left\| \frac{f^t - f}{t} + f' \right\|_1 \le \frac{|t|}{2} \|f''\|_1 \to 0, \quad (t \to 0). \qquad \square$$

We denote U_f by $\int_{\mathbb{R}} f(t)U(t)dt$.

Problem 10.4. Let A be a self-adjoint operator, and let $U(t) = e^{itA}$. Show that for every $f \in L^1(\mathbb{R})$, we have $U_f = \hat{f}(-A)$, where $\hat{f}(\xi) = \int_{\mathbb{R}} f(t)e^{-it\xi}dt$.

Theorem 10.11 (Stone theorem). *For every strongly continuous one-parameter unitary group* $U = \{U(t)\}_{t \in \mathbb{R}}$ *on* $\mathcal{H}$, *if we define an operator* A *by*

$$\mathcal{D}(A) = \left\{ x \in \mathcal{H}; \; \exists \lim_{t \to 0} \frac{1}{it}(U(t)x - x) \right\},$$

and

$$Ax = \lim_{t \to 0} \frac{1}{it}(U(t)x - x),$$

then A *is a self-adjoint operator, and* $U(t) = e^{itA}$ *holds.*

Proof. The previous lemma shows that $\mathcal{D}(A)$ is dense in $\mathcal{H}$, and

$$\left\langle \frac{1}{it}(U(t)x - x), y \right\rangle = \left\langle x, \frac{1}{-it}(U(-t)y - y) \right\rangle$$

shows that A is a symmetric operator. Note that $U(s)U(t) = U(t)U(s)$ implies $U(s)A \subset AU(s)$. We first show that A is essentially self-adjoint. For this, it suffices to show $\ker(A^* \pm iI) = \{0\}$. Let $y \in \ker(A^* \pm iI)$. Then, for every $x \in \mathcal{D}(A)$, we have

$$\frac{d}{dt}\langle U(t)x, y \rangle = i\langle AU(t)x, y \rangle = i\langle U(t)x, \mp iy \rangle = \mp \langle U(t)x, y \rangle$$

and $\langle U(t)x, y \rangle = e^{\mp t}\langle x, y \rangle$. On the other hand, since $|\langle U(t)x, y \rangle| \le \|x\|\|y\|$, we get $\langle x, y \rangle = 0$. Since $\mathcal{D}(A)$ is dense in $\mathcal{H}$, we have $y = 0$.

Note that we have $\overline{A} = A^*$. For $x \in \mathcal{D}(A)$ and $y \in \mathcal{D}(A^*)$, we have

$$\frac{d}{dt}\langle U(t)x, e^{itA^*}y \rangle = \langle iAU(t)x, e^{itA^*}y \rangle + \langle U(t)x, iA^*e^{itA^*}y \rangle.$$

Since $U(t)x \in \mathcal{D}(A)$, the right-hand side is zero. Thus, we get $U(t) = e^{itA^*}$, and Lemma 10.12 implies $A = A^*$. $\qquad\square$

We call iA the generator of $\{U(t)\}_{t \in \mathbb{R}}$.

Exercises

Exercise 10.1
Let D_c be the operator in Problem 10.1.

(1) Find $\sigma_p(D_c)$.
(2) Show $\sigma(D_c) = \sigma_p(D_c)$, and show $\sigma(D_c) = \emptyset$ for $c = 0, \infty$.

Exercise 10.2
Let L be the operator in Example 10.4.

(1) Show $\mathcal{D}(L^*) = W^{2,2}(0,1)$ and $L^* f = -f''$.
(2) Show that the deficiency indices of L are $(2,2)$.
(3) We define a linear map $\Phi : W^{2,2}(0,1) \to \mathbb{C}^4$ by $\Phi(f) = (f(0), f'(0), f(1), f'(1))$ and a sesquilinear form q on $\mathbb{C}^4$ by

$$q(v,w) = -v_1\overline{w_2} + v_2\overline{w_1} + v_3\overline{w_4} - v_4\overline{w_3}.$$

Show that $\langle L^*f, g \rangle - \langle f, L^*g \rangle = q(\Phi(f), \Phi(g))$ holds for all $f, g \in W^{2,2}(0,1)$. Show that Φ is a surjection and $\ker \Phi = \mathcal{D}(L^{**})$.
(4) Show that there exists a one-to-one correspondence between the set of self-adjoint extensions of L and the set of two-dimensional subspaces of $\mathbb{C}^4$ on which q vanishes.
(5) Show that if L' is a self-adjoint extension of L, we have $\sigma(L') = \sigma_p(L')$, and $(\lambda I - L')^{-1}$ is a compact operator for every $\lambda \in \rho(L')$.

Exercise 10.3
For $\alpha, \beta \in [0, \pi)$, we define an operator $L_{\alpha,\beta}$ on $L^2(0,1)$ whose domain $\mathcal{D}(L_{\alpha,\beta})$ is the set of $f \in W^{2,2}(0,1)$ satisfying

$$\cos\alpha f(0) + \sin\alpha f'(0) = \cos\beta f(1) + \sin\beta f'(1) = 0,$$

by $L_{\alpha,\beta} f = -f''$.

(1) Show that $L_{\alpha,\beta}$ is a self-adjoint extension of L.
(2) Show $0 \in \sigma_p(L_{\alpha,\beta}) \iff \cos\alpha\cos\beta + \cos\alpha\sin\beta - \sin\alpha\cos\beta = 0$.
(3) Show that $L_{\pi/2, 3\pi/4}$ has a negative eigenvalue.

Exercise 10.4

For $\xi, \eta \in \mathbb{R}$ and $\zeta \in \mathbb{C}$, we define $A_{\xi,\eta,\zeta} \in \mathbf{K}(L^2[0,1])_{\mathrm{sa}}$ by

$$A_{\xi,\eta,\zeta}g(t) = \int_0^t (\xi + \zeta t + (\bar{\zeta} + 1)s + \eta st)g(s)ds$$

$$+ \int_t^1 (\xi + \bar{\zeta}s + (\zeta + 1)t + \eta st)g(s)ds.$$

(1) Show that there exists a self-adjoint extension $L_{\xi,\eta,\zeta}$ of L satisfying $L_{\xi,\eta,\zeta}^{-1} = A_{\xi,\eta,\zeta}$.
(2) For $c \neq 1$, determine (ξ, η, ζ), with $D_c^* D_c = L_{\xi,\eta,\zeta}$.
(3) If $0 \notin \sigma_{\mathrm{p}}(L_{\alpha,\beta})$, determine (ξ, η, ζ), with $L_{\alpha,\beta} = L_{\xi,\eta,\zeta}$.

Exercise 10.5

Let $\tilde{D}$ be the operator in Example 10.3.

(1) Show $\{z \in \mathbb{C};\ \mathrm{Im}\, z < 0\} \subset \sigma_{\mathrm{r}}(\tilde{D}^{**})$.
(2) Show $\{z \in \mathbb{C};\ \mathrm{Im}\, z > 0\} \subset \rho(\tilde{D}^{**})$, and give an integral expression for $(zI - \tilde{D}^{**})^{-1}g$ for $z \in \rho(\tilde{D}^{**})$.
(3) Show $\sigma(\tilde{D}^{**}) = \{z \in \mathbb{C};\ \mathrm{Im}\, z \leq 0\}$.
(4) For $t \geq 0$, we define an isometry of $L^2(0, \infty)$ by

$$V_t f(s) = \begin{cases} 0, & s \leq t, \\ f(s-t), & s > t. \end{cases}$$

For $\mathrm{Re}\, z > 0$ and $f, g \in L^2(0, \infty)$, show

$$\int_0^\infty e^{-tz} \langle V_t f, g\rangle dt = \langle (zI + i\tilde{D}^{**})^{-1}f, g\rangle.$$

Exercise 10.6

Let T be a closed operator from $\mathcal{H}_1$ to $\mathcal{H}_2$ with a dense domain.

(1) Show that $T(I + T^*T)^{-1}$ is bounded.
(2) Show $(I + T^*T)^{-1}T^* \subset T^*(I + TT^*)^{-1}$.
(3) Show that the projection $\mathcal{E}_T$ from $\mathcal{H}_1 \oplus \mathcal{H}_2$ onto the graph $\mathcal{G}(T)$ of T is given by

$$\mathcal{E}_T = \begin{pmatrix} (I + T^*T)^{-1} & T^*(I + TT^*)^{-1} \\ T(I + T^*T)^{-1} & TT^*(I + TT^*)^{-1} \end{pmatrix}.$$

Exercise 10.7

Let $A = \int_{\mathbb{R}} t dE_t^A$ be the spectral decomposition of a self-adjoint operator A on $\mathcal{H}$. For $z \in \rho(A)$, let $R(z) = (zI - A)^{-1}$. Show the following Stone's formula:

$$\lim_{\varepsilon \to +0} \frac{1}{2\pi i} \int_{-\infty}^{t} \langle (R(s - i\varepsilon) - R(s + i\varepsilon)) x, x \rangle ds = \frac{1}{2} \langle (E_t^A + E_{t-0}^A) x, x \rangle.$$

Exercise 10.8

Show that the following hold for a positive self-adjoint operator A on $\mathcal{H}$:

(1) s-$\lim_{\varepsilon \to +0} (I + \varepsilon A)^{-1} = I$.
(2) s-$\lim_{t \to \infty} e^{-tA} = P$. Here, P is the projection onto $\ker A$.

Exercise 10.9

Let $A = \int_{\mathbb{R}} t dE_t^A$ be the spectral decomposition of a self-adjoint operator A on $\mathcal{H}$, and let $T \in \mathbf{B}(\mathcal{H})$. Show that the following conditions are equivalent:

(1) $TA \subset AT$.
(2) $(zI - A)^{-1}$ commutes with T for every $z \in \rho(A)$.
(3) The Cayley transform V_A of A commutes with T.
(4) E_t^A commutes with T for every $t \in \mathbb{R}$.
(5) e^{itA} commutes with T for every $t \in \mathbb{R}$.

When the above equivalent conditions hold, we say that A commutes with T.

Appendix A

Miscellaneous Facts in Analysis

A.1   Convergence of a Net

We say that an ordered set Λ is *directed* if every finite subset of Λ has an upper bound.

Example A.1. The set of finite subsets $\mathfrak{F}_X$ of a set X is a directed set by the inclusion relation.

Example A.2. Let X be a topological space and let $x \in X$. The set of neighborhoods $\mathfrak{N}(x)$ of x is a directed set by defining $V \leq U$ if $U \subset V$.

We call a map $i : \Lambda \to X$ from a directed set Λ to a set X a *net*. As in the case of sequences, it is customary to write $i(\lambda) = x_\lambda$ and express a net as $\{x_\lambda\}_{\lambda \in \Lambda}$.

We say that a net $\{x_\lambda\}_{\lambda \in \Lambda}$ in a topological space X converges to $x \in X$ if the following condition holds: for every $U \in \mathfrak{N}(x)$, there exists $\lambda_0 \in \Lambda$ such that $\lambda_0 \leq \lambda$ implies $x_\lambda \in U$. When this condition holds, we say that x is the limit of the net $\{x_\lambda\}_{\lambda \in \Lambda}$. If X is a Hausdorff space, the limit of a net is unique if it exists, and we write

$$\lim_{\lambda \in \Lambda} x_\lambda = x.$$

It is known that giving a topology is equivalent to determining the convergence of nets by axiomatizing the notion of the convergence of nets (see Kelley, 1975, Chapter 2). In a nutshell, every topological information can be, in principle, encoded in the convergence of nets.

Lemma A.1. *Let X be a topological space and let A be a subset of X. Then, for $x \in X$, the following conditions are equivalent*:

(1) $x \in \overline{A}$.
(2) *There exists a net $\{a_\lambda\}_{\lambda \in \Lambda}$ in A converging to x.*

Proof. (1) $\implies$ (2). For each $U \in \mathfrak{N}(x)$, we choose $a_U \in U \cap A$. Then, $\{a_U\}_{U \in \mathfrak{N}(x)}$ is a net in A converging to x.

(2) $\implies$ (1). Let $\{a_\lambda\}_{\lambda \in \Lambda}$ be a net in A converging to x. For every $U \in \mathfrak{N}(x)$, there exists $\lambda_0 \in \Lambda$ such that $\lambda_0 \leq \lambda$ implies $a_\lambda \in U$, and in particular $A \cap U \neq \emptyset$. Thus, $x \in \overline{A}$. $\square$

Problem A.1. Let X and Y be topological spaces.

(1) Show that a subset A of X is closed if and only if the following holds: if $\{a_\lambda\}_{\lambda \in \Lambda}$ is a net in A converging in X, then its limit belongs to A.
(2) Show that a map $f : X \to Y$ is continuous at $a \in X$ if and only if the following holds: if $\{x_\lambda\}_{\lambda \in \Lambda}$ is a net in X converging to a, then the net $\{f(x_\lambda)\}_{\lambda \in \Lambda}$ converges to $f(a)$.

When we characterize compactness by the convergence of nets, we should be careful about the definition of a subnet of a net. Let $\{x_\lambda\}_{\lambda \in \Lambda}$ be a net in a set X. Let M be a directed set, and assume that a map $\varphi : M \to \Lambda$ satisfies the following condition: $\forall \lambda \in \Lambda$, $\exists \mu_0 \in M$, $\forall \mu \geq \mu_0$, $\lambda \leq \varphi(\mu)$. When this condition holds, we call $\{x_{\varphi(\mu)}\}_{\mu \in M}$ a *subnet* of $\{x_\lambda\}_{\lambda \in \Lambda}$. Note that we do not assume that φ is order-preserving.

Proposition A.1. *For a topological space X, the following conditions are equivalent*:

(1) *X is compact.*
(2) *Every net in X has a convergent subnet.*

Proof. (1) $\implies$ (2) Let $\{x_\lambda\}_{\lambda \in \Lambda}$ be a net in X. For $\lambda \in \Lambda$, we denote by F_λ the closure of $\{x_\mu; \lambda \leq \mu\}$. Then, since $\{F_\lambda\}_{\lambda \in \Lambda}$ is a family of closed sets with the finite intersection property, there exists $x \in \bigcap_{\lambda \in \Lambda} F_\lambda$. For $(U, \lambda), (V, \mu) \in \mathfrak{N}(x) \times \Lambda$, we set $(U, \lambda) \leq (V, \mu)$ if $U \leq V$ and $\lambda \leq \mu$. Then, $\mathfrak{N}(x) \times \Lambda$ is a directed set. Since $x \in F_\lambda$ holds for every $\lambda \in \Lambda$, we get $U \cap \{x_\mu; \lambda \leq \mu\} \neq \emptyset$ for every $U \in \mathfrak{N}(x)$, and we can choose μ satisfying $x_\mu \in U$ and $\lambda \leq \mu$. We fix such μ and set $\varphi(U, \lambda) = \mu$. Since $\lambda \leq \varphi(U, \lambda)$, the net $\{x_{\varphi(U,\lambda)}\}_{(U,\lambda)\in\mathfrak{N}(x)\times\Lambda}$ is a subnet of $\{x_\lambda\}_{\lambda\in\Lambda}$. As $x_{\varphi(U,\lambda)} \in U$, it converges to x.

(2) $\implies$ (1). Let $\{Y_i\}_{i\in I}$ be a family of closed subsets of X with the finite intersection property. Then, for every $F \in \mathfrak{F}_I$, we have $\bigcap_{i\in F} Y_i \neq \emptyset$, and we choose an element in this set and denote it by x_F. From the assumption, there exists a subnet $\{x_{\varphi(\mu)}\}_{\mu\in M}$ of $\{x_F\}_{F\in\mathfrak{F}_\Lambda}$ converging to an element $x \in X$. From the property of φ, the following holds: $\forall i_0 \in I, \exists \mu_0 \in M, \forall \mu \geq \mu_0, \{i_0\} \leq \varphi(\mu)$. This means that $i_0 \in \varphi(\mu)$ holds for all $\mu \geq \mu_0$, and $x_{\varphi(\mu)} \in Y_{i_0}$. Since the net $\{x_{\varphi(\mu)}\}_{\mu_0\leq\mu}$ also converges to x, we get $x \in Y_{i_0}$ as Y_{i_0} is closed. Since $i_0 \in I$ is arbitrary, we get $x \in \bigcap_{i\in I} Y_i$. $\qquad\square$

Assume that the topology of a locally convex space X is given by a family of semi-norms $\{p_i\}_{i\in I}$. We say that a net $\{x_\lambda\}_{\lambda\in\Lambda}$ in X is a *Cauchy net* if the following holds: $\forall \varepsilon > 0, \forall F \in I, \exists \lambda_0 \in \Lambda$, $\forall \lambda, \mu \geq \lambda_0, \forall i \in F, p_i(x_\lambda - x_\mu) < \varepsilon$. If every Cauchy net in X converges, we say that X is *complete*.

Every Cauchy net $\{x_\lambda\}_{\lambda\in\Lambda}$ in a Banach space converges. Indeed, from the definition of a Cauchy net, the following holds: $\forall n \in \mathbb{N}$, $\exists \lambda_n \in \Lambda, \forall \lambda, \mu \geq \lambda_n, \|x_\lambda - x_\mu\| < 1/n$. Furthermore, we can inductively take $\{\lambda_n\}_{n=1}^\infty$ so that $n < m$ implies $\lambda_n \leq \lambda_m$. Then, $\{x_{\lambda_n}\}_{n=1}^\infty$ is a Cauchy sequence, and it converges to some x. For $\lambda \geq \lambda_n$, we have $\|x-x_\lambda\| \leq \|x-x_{\lambda_n}\| + \|x_{\lambda_n} - x_\lambda\| < 2/n$, and $\{x_\lambda\}_{\lambda\in\Lambda}$ converges to x as well.

A.2 Basics of L^p Spaces

In this section, we prove the basic properties of L^p spaces. In what follows, (Ω, μ) means a measure space.

Problem A.2. Let $1 < p, q < \infty$ with $1/p + 1/q = 1$. Show the inequality

$$ab \leq \frac{a^p}{p} + \frac{b^q}{q}$$

holds for all $a, b > 0$, and show that the equality holds if and only if $a^p = b^q$.

Theorem A.1. *Let $1 \leq p, q \leq \infty$, with $1/p + 1/q = 1$. Then, the following inequalities hold:*

(1) *Hölder inequality:*

$$\forall f \in L^p(\Omega, \mu), \ \forall g \in L^q(\Omega, \mu), \quad \left| \int_\Omega fg d\mu \right| \leq \|f\|_p \|g\|_q.$$

(2) *Minkowski inequality:*

$$\forall f, g \in L^p(\Omega, \mu), \quad \|f + g\|_p \leq \|f\|_p + \|g\|_p.$$

Proof. Since the cases of $p = 1, \infty$ are easy, we show the statement for $1 < p, q < \infty$.

(1) Letting $a = \|f\|_p^{-1}|f|$ and $b = \|g\|_q^{-1}|g|$ in the inequality in the above problem and integrating both sides, we get

$$\frac{1}{\|f\|_p \|g\|_q} \int_\Omega |fg| d\mu \leq \int_\Omega \left(\frac{|f|^p}{p\|f\|_p^p} + \frac{|g|^q}{q\|g\|_q^q} \right) d\mu = 1,$$

and the Hölder inequality holds.

(2) Note that for $h \in L^p(\Omega, \mu)$, we have $|h|^{p-1} = |h|^{\frac{p}{q}} \in L^q(\Omega, \mu)$. The triangle inequality and the Hölder inequality imply

$$\begin{aligned} \|f + g\|_p^p &= \int_\Omega |f + g||f + g|^{p-1} d\mu \\ &\leq \int_\Omega |f||f + g|^{\frac{p}{q}} d\mu + \int_\Omega |g||f + g|^{\frac{p}{q}} d\mu \\ &\leq \|f\|_p \|f + g\|_p^{\frac{p}{q}} + \|g\|_p \|f + g\|_p^{\frac{p}{q}} \\ &= (\|f\|_p + \|g\|_q) \|f + g\|_p^{p-1}, \end{aligned}$$

and the Minkowski inequality holds. $\square$

Theorem A.2. *For $1 \leq p \leq \infty$, the function space $(L^p(\Omega, \mu), \| \cdot \|)$ is a Banach space.*

Proof. The Minkowski inequality shows that $(L^p(\Omega, \mu), \| \cdot \|)$ is a normed space, and it suffices to show completeness. We leave the proof in the case of $p = \infty$ to the reader, and we show completeness in the case of $1 \leq p < \infty$. Let $\{f_n\}_{n=1}^{\infty}$ be a Cauchy sequence in $L^p(\Omega, \mu)$. Since a Cauchy sequence with a convergent subsequence is convergent, we may assume $\|f_{n+1} - f_n\|_p < 2^{-n}$ by taking a subsequence if necessary. Let

$$g_n = |f_1| + \sum_{k=1}^{n-1} |f_{k+1} - f_k|.$$

Since $\{g_n(\omega)\}_{n=1}^{\infty}$ is monotone increasing, there exists a pointwise limit $g : \Omega \to [0, \infty]$. Since

$$\|g_n\|_p \leq \|f_1\|_p + \sum_{k=1}^{n-1} \|f_{k+1} - f_k\|_p \leq \|f_1\|_p + 1,$$

the monotone convergence theorem implies

$$\int_{\Omega} |g|^p d\mu = \lim_{n \to \infty} \int_{\Omega} |g_n|^p d\mu \leq (\|f_1\|_p + 1)^p.$$

Thus, g is finite almost everywhere, and $g \in L^p(\Omega, \mu)$. From this,

$$f(\omega) = f_1(\omega) + \sum_{k=1}^{\infty} (f_{k+1}(\omega) - f_k(\omega)) \quad \left(= \lim_{n \to \infty} f_n(\omega) \right)$$

absolutely converges almost everywhere, and $|f(\omega)| \leq |g(\omega)|$ holds. Since $|f(\omega) - f_n(\omega)|^p \leq |g(\omega)|^p$, the Lebesgue theorem implies

$$\lim_{n \to \infty} \int_{\Omega} |f(\omega) - f_n(\omega)|^p d\mu = \int_{\Omega} \lim_{n \to \infty} |f(\omega) - f_n(\omega)|^p d\mu = 0,$$

and $\{f_n\}_{n=1}^{\infty}$ converges to f in $L^p(\Omega, \mu)$.   □

Corollary A.1. *Let $1 \leq p \leq \infty$, let $f \in L^p(\Omega, \mu)$, and let $\{f_n\}_{n=1}^{\infty}$ be a sequence in $L^p(\Omega, \mu)$. If $\lim_{n \to \infty} \|f_n - f\|_p = 0$, there exists a subsequence of $\{f_n\}_{n=1}^{\infty}$ converging to f almost everywhere.*

Proof.   Again, we treat only the case of $1 \leq p < \infty$. The proof of the previous theorem shows that if we take a subsequence $\{f_{n_k}\}_{k=1}^{\infty}$

of $\{f_n\}_{n=1}^{\infty}$ such that $\|f_{n_{k+1}} - f_{n_k}\|_p < 2^{-k}$ holds for all $k \in \mathbb{N}$, it satisfies the condition. □

A.3 L^p–L^q Duality

The standard proof of Theorem 3.4 is a measure-theoretic argument using the Radon–Nikodym theorem (Conway, 1990, Appendix B; Rudin, 1987, Theorem 6.16). However, as L^p for $p = 2$ is a Hilbert space, Theorem 3.4 follows from the Riesz representation theorem, whose proof is a geometric argument using orthogonal decomposition. Also, for $p = 1$, there is an argument reducing the proof to the Riesz representation theorem (see Exercise 3.2). Here, we give a proof of Theorem 3.4 in the case of $1 < p < \infty$ by using a geometric argument without assuming σ-finiteness of the measure space (Ω, μ). The point of the proof is that we first show the reflexivity of $L^p(\Omega, \mu)$ for $2 \leq p < \infty$. We assume $1 < p, q < \infty$ and $1/p + 1/q = 1$ throughout this section.

We first introduce a notion to capture a geometric property of spaces resembling a Hilbert space such as the L^p spaces.

Definition A.1. We say that a Banach space X is *uniformly convex* if the following holds: $\forall \varepsilon > 0$, $\exists \delta = \delta(\varepsilon) > 0$, $x, y \in B_X$, $\|x - y\| \geq \varepsilon \implies \|2^{-1}(x + y)\| \leq 1 - \delta$.

The uniform convexity of X means that we can quantitatively estimate how far the midpoint of two points on the unit sphere lies from the unit sphere in terms of the distance between the two points.

Example A.3. A Hilbert space is uniformly convex by $\delta(\varepsilon) = 1 - \sqrt{1 - \varepsilon^2/4}$.

Theorem A.3. *Every uniformly convex space X is reflexive.*

Proof.   Let $x^{**} \in X^{**}$, with $\|x^{**}\| = 1$. Since B_X is dense in $B_{X^{**}}$ in the weak* topology $\sigma(X^{**}, X^*)$ thanks to Corollary 5.4, there exists a net $\{x_\lambda\}_{\lambda \in \Lambda}$ in B_X converging to x^{**} in the weak* topology. We first show that the following holds: $\forall \delta > 0$, $\exists \lambda_0 \in \Lambda$, $\forall \lambda, \mu \geq \lambda_0$, and $\|2^{-1}(x_\lambda + x_\mu)\| \geq 1 - \delta$. Since $\|x^{**}\| = 1$, there exists $\varphi \in B_{X^*}$ satisfying $x^{**}(\varphi) > 1 - \delta/2$. Since $\{x_\lambda\}_{\lambda \in \Lambda}$ converges to x^{**} in the topology $\sigma(X^{**}, X^*)$, there exists $\lambda_0 \in \Lambda$ such that $\lambda \geq \lambda_0$ implies $|x^{**}(\varphi) - \varphi(x_\lambda)| < \delta/2$. Thus, for $\lambda, \mu \geq \lambda_0$,

$$\left| \varphi \left(\frac{1}{2}(x_\lambda + x_\mu) \right) \right| > |x^{**}(\varphi)| - \frac{|x^{**}(\varphi) - \varphi(x_\lambda)| + |x^{**}(\varphi) - \varphi(x_\mu)|}{2}$$

$$> 1 - \delta.$$

Since $\varphi \in B_{X^*}$, we get $\|2^{-1}(x_\lambda + x_\mu)\| > 1 - \delta$.

Using the uniform convexity of X, we can apply the above fact to $\delta(\varepsilon)$ and get the following: $\forall \varepsilon > 0$, $\exists \lambda_0 \in \Lambda$, $\forall \lambda, \mu \geq \lambda_0$, and $\|x_\lambda - x_\mu\| < \varepsilon$. Thus, $\{x_\lambda\}_{\lambda \in \Lambda}$ is a Cauchy net and converges to some $x \in B_X$ in norm. Since the limit in the norm convergence coincides with the limit in the weak* convergence, we get $x^{**} = x$, which shows $X^{**} = X$. $\qquad\square$

Next, to show the uniform convexity of $L^p(\Omega, \mu)$ for $2 \leq p < \infty$, we prove the Clarkson inequality.

Problem A.3. For $1 \leq r < p \leq \infty$ and $a \in \mathbb{C}^n$, show the following:

$$\|a\|_p \leq \|a\|_r \leq n^{\frac{1}{r} - \frac{1}{p}} \|a\|_p.$$

Theorem A.4 (Clarkson inequality). *For all $2 \leq p < \infty$ and $f, g \in L^p(\Omega, \mu)$, the inequality*

$$\left\| \frac{1}{2}(f + g) \right\|_p^p + \left\| \frac{1}{2}(f - g) \right\|_p^p \leq \frac{1}{2}(\|f\|_p^p + \|g\|_p^p),$$

holds. In particular, $L^p(\Omega, \mu)$ is uniformly convex.

Proof. From the case of $n = 2$ and $r = 2$ in the above problem, for $a, b \in \mathbb{C}$,

$$(|a + b|^p + |a - b|^p)^{\frac{1}{p}} \leq (|a + b|^2 + |a - b|^2)^{\frac{1}{2}}$$

$$= 2^{\frac{1}{2}}(|a|^2 + |b|^2)^{\frac{1}{2}}$$

$$\leq 2^{1 - \frac{1}{p}}(|a|^p + |b|^p)^{\frac{1}{p}}$$

holds, and

$$|f(\omega) + g(\omega)|^p + |f(\omega) - g(\omega)|^p \le 2^{p-1}(|f(\omega)|^p + |g(\omega)|^p).$$

Integrating both sides, we get the Clarkson inequality.

For $f, g \in B_{L^p(\Omega,\mu)}$,

$$\left\|\frac{1}{2}(f+g)\right\|_p \le \left(1 - \left\|\frac{1}{2}(f-g)\right\|_p^p\right)^{\frac{1}{p}}$$

holds, and $L^p(\Omega, \mu)$ is uniformly convex, with $\delta(\varepsilon) = 1 - (1 - 2^{-p}\varepsilon^p)^{1/p}$. $\qquad\square$

Remark A.1. It is known that for all $1 < p \le 2$ and $f, g \in L^p(\Omega, \mu)$,

$$\left\|\frac{1}{2}(f+g)\right\|_p^q + \left\|\frac{1}{2}(f-g)\right\|_p^q \le \left(\frac{\|f\|_p^p + \|g\|_p^p}{2}\right)^{\frac{q}{p}}$$

holds, and it is also called the Clarkson inequality (see Carothers, 2005, Theorem 11.15). From this, we can see that $L^p(\Omega, \mu)$ is uniformly convex for $1 < p < 2$ as well, although we do not need it for our purpose, and we omit its proof.

Proof of Theorem 3.4. We first show the statement for $2 \le p < \infty$. Since Φ is an isometry, the image of Φ is a closed subspace of $L^p(\Omega, \mu)^*$. If $L^p(\Omega, \mu)^* \ne \Phi(L^q(\Omega, \mu))$, we would have $\Phi(L^q(\Omega, \mu))^\perp \ne \{0\}$. Since $L^p(\Omega, \mu)$ is reflexive, we get

$$\Phi(L^q(\Omega, \mu))^\perp = \{f \in L^p(\Omega, \mu); \ \forall g \in L^q(\Omega, \mu), \ \varphi_g(f) = 0\}.$$

On the other hand, if we put for $f \in L^p(\Omega, \mu)$

$$g(\omega) = \begin{cases} |f(\omega)|^{\frac{p}{q}-1}\overline{f(\omega)}, & f(\omega) \ne 0, \\ 0, & f(\omega) = 0, \end{cases}$$

we get $g \in L^q(\Omega, \mu)$ and $\varphi_g(f) = \|f\|_p^p$. Thus, $\Phi(L^q(\Omega, \mu))^\perp = \{0\}$, and Φ is a surjection.

For $1 < p < 2$, we have $2 < q < \infty$, and the above argument shows that through the bilinear form

$$L^p(\Omega, \mu) \times L^q(\Omega, \mu) \ni (f, g) \mapsto \int_\Omega fg\, d\mu,$$

$L^p(\Omega, \mu)$ is identified with the dual space of $L^q(\Omega, \mu)$. Since $L^q(\Omega, \mu)$ is reflexive, the map Φ is a surjection. $\qquad\qquad\qquad\square$

A.4 Stone–Weierstrass Theorem

In this section, we assume that Ω is a compact Hausdorff space unless otherwise stated. The Stone–Weierstrass theorem is a generalization of the Weierstrass polynomial approximation theorem, and it is a very useful theorem for giving a criterion for a subalgebra of $C(\Omega)$ to be dense in $C(\Omega)$.

Let $\mathcal{A}$ be a subset of $C(\Omega)$. We say that $\mathcal{A}$ separates any two points in Ω if, for any two different points $\omega_1, \omega_2 \in \Omega$, there exists $f \in \mathcal{A}$ satisfying $f(\omega_1) \neq f(\omega_2)$.

To show the theorem in the case of real-valued functions, we denote by $C_{\mathbb{R}}(\Omega)$ the set of real-valued continuous functions on Ω. The space $C_{\mathbb{R}}(\Omega)$ is a real Banach space with $\|\cdot\|_\infty$. In what follows, we say that $\mathcal{A}$ is a subalgebra of $C(\Omega)$ if it is a subalgebra over $\mathbb{C}$, and say that $\mathcal{A}$ is a subalgebra of $C_{\mathbb{R}}(\Omega)$ if it is a subalgebra over $\mathbb{R}$.

For $f, g \in C_{\mathbb{R}}(\Omega)$, we set $f \vee g(\omega) = \max\{f(\omega), g(\omega)\}$, $f \wedge g(\omega) = \min\{f(\omega), g(\omega)\}$.

Lemma A.2. *Let $\mathcal{A}$ be a closed subalgebra of $C_{\mathbb{R}}(\Omega)$. Then, $f, g \in \mathcal{A}$ implies $f \vee g$ and $f \wedge g \in \mathcal{A}$.*

Proof. Since $f \vee g = 2^{-1}(|f - g| + f + g)$, $f \wedge g = -((-f) \vee (-g))$, it suffices to show that $f \in \mathcal{A}$ implies $|f| \in \mathcal{A}$. Furthermore, it suffices to show it for $\|f\|_\infty \leq 1$. Since we have

$$\left| \sqrt{\varepsilon^2 + t^2} - |t| \right| = \frac{\varepsilon^2}{\sqrt{\varepsilon^2 + t^2} + |t|} \leq \varepsilon,$$

for $\varepsilon > 0$, it suffices to show that $h(t) = \sqrt{\varepsilon^2 + t}$ can be approximated by polynomials with real coefficients uniformly on $[0, 1]$. Indeed, if a polynomial $p(t)$ with real coefficients approximates h on $[0, 1]$ within ε, the polynomial $q(t) = p(t^2) - p(0)$ has no constant term, and

$$|q(t) - |t|| \leq |p(0)| + |p(t^2) - \sqrt{\varepsilon^2 + t^2}| + \left| \sqrt{\varepsilon^2 + t^2} - |t| \right| < 4\varepsilon,$$

for all $t \in [-1, 1]$, and we get $q(f) \in \mathcal{A}$ and $\|q(f) - |f|\|_\infty < 4\varepsilon$.

Since the function $h(z) = \sqrt{\varepsilon^2 + z}$ is holomorphic on the disk centered at $z = 1/2$ with radius $1/2 + \varepsilon^2$, the Taylor series

$$h(t) = \sum_{n=0}^{\infty} \frac{h^{(n)}(\frac{1}{2})}{n!} \left(t - \frac{1}{2} \right)^n$$

has the radius of convergence strictly larger than $1/2$, and this series uniformly converges on $[0, 1]$. Thus, an appropriate partial sum gives the desired approximation. $\quad\square$

Theorem A.5 (Stone–Weierstrass theorem (real case)). *Let $\mathcal{A}$ be a subalgebra of $C_{\mathbb{R}}(\Omega)$ separating any two points in Ω and containing the constant function 1. Then, $\mathcal{A}$ is dense in $C_{\mathbb{R}}(\Omega)$.*

Proof. Since the closure of $\mathcal{A}$ satisfies the same assumption, it suffices to show $\mathcal{A} = C_{\mathbb{R}}(\Omega)$, assuming that $\mathcal{A}$ is closed from the beginning.

Let $f \in C_{\mathbb{R}}(\Omega)$. We first show that for any $\varepsilon > 0$ and $\xi \in \Omega$, there exists $g_{\xi} \in \mathcal{A}$ such that $f \le g_{\xi} + \varepsilon$ and $f(\xi) = g_{\xi}(\xi)$. Indeed, since $\mathcal{A}$ is a real subspace separating any two points in Ω and containing 1, for any two distinct points ξ, η in Ω, there exists $g_{\xi,\eta} \in \mathcal{A}$ satisfying $f(\xi) = g_{\xi,\eta}(\xi)$ and $f(\eta) = g_{\xi,\eta}(\eta)$. The set $V_{\eta} = \{\omega \in \Omega;\ f(\omega) < g_{\xi,\eta}(\omega) + \varepsilon\}$ is an open neighborhood of η. Since Ω is compact, there exist $\eta_1, \eta_2, \ldots, \eta_n \in \Omega$ satisfying $\Omega = \bigcup_{i=1}^{n} V_{\eta_i}$. Let $g_{\xi} = g_{\xi,\eta_1} \vee g_{\xi,\eta_2} \vee \cdots \vee g_{\xi,\eta_n}$. Then, g_{ξ} satisfies the condition.

For each $\xi \in \Omega$, we set $U_{\xi} = \{\omega \in \Omega;\ g_{\xi}(\omega) - \varepsilon < f(\omega)\}$. Then, U_{ξ} is an open neighborhood of ξ. Thus, there exist $\xi_1, \xi_2, \ldots, \xi_m \in \Omega$ satisfying $\Omega = \bigcup_{i=1}^{m} U_{\xi_i}$. Let $g = g_{\xi_1} \wedge g_{\xi_2} \wedge \cdots \wedge g_{\xi_m}$. Then, as $g - \varepsilon \le f \le g + \varepsilon$ holds, we get $\|f - g\|_{\infty} \le \varepsilon$. $\quad\square$

Corollary A.2 (Weierstrass polynomial approximation theorem). *Every real-valued continuous function on a finite closed interval is uniformly approximated by polynomials with real coefficients.*

Theorem A.6 (Stone–Weierstrass theorem (complex case)). *Let $\mathcal{A}$ be a subalgebra of $C(\Omega)$ satisfying the following conditions:*

(1) $\mathcal{A}$ *separates any two points in Ω.*
(2) $1 \in \mathcal{A}$.
(3) $f \in \mathcal{A}$ *implies $\overline{f} \in \mathcal{A}$.*

Then, $\mathcal{A}$ is dense in $C(\Omega)$.

Proof. The space $\mathcal{A}_\mathbb{R} = \mathcal{A} \cap C_\mathbb{R}(\Omega)$ is a subalgebra of $C_\mathbb{R}(\Omega)$, and condition (3) implies $\mathcal{A} = \mathcal{A}_\mathbb{R} + i\mathcal{A}_\mathbb{R}$. Since $\mathcal{A}_\mathbb{R}$ separates any two points in Ω and $1 \in \mathcal{A}_\mathbb{R}$, the previous theorem shows that $\mathcal{A}_\mathbb{R}$ is dense in $C_\mathbb{R}(\Omega)$, and $\mathcal{A}$ is dense in $C(\Omega)$. □

Corollary A.3. *The set of trigonometric polynomials is dense in* $C(\mathbb{T})$.

Theorem A.7. *Let Ω be a compact Hausdorff space. Then, the following conditions are equivalent*:

(1) Ω *is metrizable.*
(2) $C(\Omega)$ *is separable.*

Proof. (1) $\Longrightarrow$ (2). We fix a metric d of Ω. Note that a compact metric space is separable. We choose a countable dense subset $\{\omega_n\}_{n=1}^\infty$ of Ω, and we set $f_0 = 1$ and $f_n(\omega) = d(\omega, \omega_n)$ for $n \in \mathbb{N}$. The Stone–Weierstrass theorem shows that the algebra generated by $\{f_n\}_{n=0}^\infty$ is dense in $C(\Omega)$, and $C(\Omega)$ is separable.

 (2) $\Longrightarrow$ (1). Since $C(\Omega)$ is separable, Problem 3.7 implies that $B_{C(\Omega)^*}$ is metrizable in the weak* topology. We define $\rho : \Omega \to B_{C(\Omega)^*}$ by $\rho(\omega)(f) = f(\omega)$. Then, ρ is injective and continuous with respect to the weak* topology. Indeed, if $\{\omega_\lambda\}_{\lambda \in \Lambda}$ is a net in Ω converging to $\omega \in \Omega$, then for every $f \in C(\Omega)$, we have

$$\lim_{\lambda \in \Lambda} \rho(\omega_\lambda)(f) = \lim_{\lambda \in \Lambda} f(\omega_\lambda) = f(\omega) = \rho(\omega)(f),$$

and $\{\rho(\omega_\lambda)\}_{\lambda \in \Lambda}$ converges to $\rho(\omega)$ in the weak* topology. Since Ω is compact, the map ρ is a homeomorphism from Ω onto $\rho(\Omega)$, and Ω is metrizable. □

 Before ending this section, we consider the case where Ω is a locally compact Hausdorff space. We denote by $\tilde{\Omega} = \Omega \cup \{\infty\}$ the one-point compactification of Ω. Let $C_0(\Omega)$ be the set of continuous functions on Ω vanishing at infinity. Then, $C_0(\Omega)$ is identified with $\{f \in C(\tilde{\Omega}); \ f(\infty) = 0\}$ (see Exercise 1.3).

Theorem A.8 (Stone–Weierstrass theorem (locally compact case)). *Let Ω be a locally compact Hausdorff space, and let $\mathcal{A}$ be a subalgebra of $C_0(\Omega)$ satisfying the following conditions:*

(1) $\mathcal{A}$ *separates any two points in* Ω.
(2) *For every* $\omega \in \Omega$, *there exists* $f \in \mathcal{A}$ *satisfying* $f(\omega) \neq 0$.
(3) $f \in \mathcal{A}$ *implies* $\overline{f} \in \mathcal{A}$.

Then, $\mathcal{A}$ *is dense in* $C_0(\Omega)$.

Proof. Identifying the constant function 1 on Ω with the constant function 1 on $\tilde{\Omega}$, we consider $C_0(\Omega), \mathbb{C}1 \subset C(\tilde{\Omega})$ in the following argument. Note that for $f \in C_0(\Omega)$ and $\alpha \in \mathbb{C}$, we can show $\|f + \alpha\|_\infty \geq |\alpha|$ by considering the value at ∞.

Let $\tilde{\mathcal{A}} = \mathcal{A} + \mathbb{C}1$. Then, $\tilde{\mathcal{A}}$ is a subalgebra of $C(\tilde{\Omega})$ satisfying the assumptions in the previous theorem and is dense in $C(\tilde{\Omega})$. Thus, for every $f \in C_0(\Omega)$, there exist $f_n \in \mathcal{A}$ and $\alpha_n \in \mathbb{C}$ such that $\{f_n + \alpha_n\}_{n=1}^\infty$ converges to f in $C(\tilde{\Omega})$. From

$$\|f_n - f\|_\infty \leq \|f_n + \alpha_n - f\|_\infty + |\alpha_n| \leq 2\|f_n + \alpha_n - f\|_\infty,$$

the sequence $\{f_n\}_{n=1}^\infty$ converges to f in $C_0(\Omega)$. $\qquad\square$

A.5 Regular Borel Measures

Here, we gather descriptions of the regularity of Borel measures that are necessary in this book. The symbol Ω means a locally compact Hausdorff space in what follows. Recall that $\mathfrak{B}_\Omega$ is the σ-algebra of all Borel subsets of Ω. We denote by $\mathfrak{K}_\Omega$ the set of compact subsets of Ω, and by $\mathfrak{O}_\Omega$ the set of open subsets of Ω.

Before discussing measures, we note basic facts on the topology of a locally compact space. For every $K \in \mathfrak{K}_\Omega$, there exists $U \in \mathfrak{O}_\Omega$ containing K such that its closure is compact. Since the one-point compactification $\tilde{\Omega} = \Omega \cup \{\infty\}$ of Ω is a normal space, and K and $\tilde{\Omega} \backslash U$ are two disjoint closed subsets of $\tilde{\Omega}$, Urysohn's lemma shows that there exists a continuous function, $\rho : \tilde{\Omega} \to [0, 1]$, such that it is 1 on K and 0 on the complement of U. Since $\mathrm{supp}\, \rho \subset \overline{U}$, we have $\rho \in C_c(\Omega)$. Also, the Tietze extension theorem shows that every $f \in C(K)$ extends to a continuous function on $\tilde{\Omega}$. Multiplying it by ρ, we see that f extends to an element in $C_c(\Omega)$.

Definition A.2. Let Ω be a locally compact Hausdorff space, and let μ be a σ-finite Borel measure on Ω such that $\mu(K) < \infty$ for all $K \in \mathfrak{K}_\Omega$. We say that μ is *inner regular* if $\mu(E) = \sup\{\mu(K);$

$K \in \mathfrak{K}_\Omega, K \subset E\}$ holds for every $E \in \mathfrak{B}_\Omega$. We say that μ is *outer regular* if $\mu(E) = \inf\{\mu(O); O \in \mathfrak{O}_\Omega, E \subset O\}$ holds for every $E \in \mathfrak{B}_\Omega$. We say that μ is *regular* if μ is inner regular and outer regular. When Ω is compact, inner regularity coincides with outer regularity.

Proposition A.2. *Let μ be a regular Borel measure on Ω satisfying the above assumption. Then, the following hold:*

(1) *There exists the largest open set O_μ satisfying $\mu(O_\mu) = 0$. We call $\Omega \backslash O_\mu$ the support of μ and denote it by $\operatorname{supp}\mu$.*
(2) *For $1 \le p < \infty$, the space $C_c(\Omega)$ is dense in $L^p(\Omega, \mu)$.*

Proof. (1) Let

$$O_\mu = \bigcup_{O \in \mathfrak{O}_\Omega, \, \mu(O)=0} O.$$

Then, O_μ is an open subset of Ω. For $K \in \mathfrak{K}_\Omega$ with $K \subset O_\mu$, there exist finitely many open sets $O_1, O_2, \ldots, O_n \in \mathfrak{O}_\Omega$ such that $\mu(O_i) = 0$ and $K \subset \bigcup_{i=1}^n O_i$. Thus, $\mu(K) = 0$, and the regularity of μ implies $\mu(O_\mu) = 0$.

(2) It suffices to show that for every $E \in \mathfrak{B}_\Omega$ with $\mu(E) < \infty$, the indicator function χ_E is approximated by functions in $C_c(\Omega)$ in the L^p-norm. For every $\varepsilon > 0$, there exists $K \in \mathfrak{K}_\Omega$ satisfying $K \subset E$ and $\mu(E \backslash K) < \varepsilon$, and so $\|\chi_E - \chi_K\|_p < \varepsilon^{1/p}$. Also, there exists $O \in \mathfrak{O}_\Omega$ satisfying $K \subset O$ and $\mu(O \backslash K) < \varepsilon$. We take $U \in \mathfrak{O}_\Omega$ containing K with compact closure and a continuous function $\rho : \Omega \to [0,1]$ such that it is 1 on K and 0 on $\Omega \backslash (O \cap U)$. Then, since $\operatorname{supp}\rho \subset \bar{U}$, we have $\rho \in C_c(\Omega)$. Since $\|\rho - \chi_K\|_p < \varepsilon^{1/p}$, we get $\|\chi_E - \rho\|_p < 2\varepsilon^{1/p}$. $\square$

When Ω is a compact metric space, every finite Borel measure on Ω is regular. Indeed, in this case, every open set is an F_σ-set (a countable union of closed sets), and the family of sets

$$\{E \in \mathfrak{B}_\Omega; \mu(E) = \sup\{\mu(K); \, K \in \mathfrak{K}_\Omega, \, K \subset E\}$$
$$= \inf\{\mu(O); \, O \in \mathfrak{O}_\Omega, \, E \subset O\}\}$$

is a σ-algebra containing all open subsets of Ω, and it coincides with $\mathfrak{B}_\Omega$. From the above proposition, the space $C(\Omega)$ is dense in $L^p(\Omega, \mu)$ for $1 \le p < \infty$. Since the embedding of $C(\Omega)$ into $L^p(\Omega, \mu)$ is continuous, the Stone–Weierstrass theorem implies that $L^p(\Omega, \mu)$

is separable. We generalize this fact to the case where the one-point compactification of Ω is metrizable, allowing σ-finite measures.

We say that a topological space is σ-compact if it is a countable union of compact subsets.

Theorem A.9. *Let Ω be a locally compact and σ-compact metric space, and let μ be a Borel measure on Ω such that μ is finite for every compact subset of Ω. Then:*

(1) *μ is regular.*
(2) *$C_c(\Omega)$ and $C_0(\Omega)$ are separable.*
(3) *$L^p(\Omega, \mu)$ is separable and $C_c(\Omega)$ is dense in $L^p(\Omega, \mu)$ for all $1 \leq p < \infty$.*

Proof. (1) Since Ω is σ-compact, there exists an increasing sequence of compact subsets $\{K_n\}_{n=1}^\infty$ of Ω whose union is Ω. Since Ω is locally compact, there exists an open set U_n containing K_n with compact closure. By inductively replacing K_{n+1} with $K_{n+1} \cup \overline{U_n}$ if necessary, we may assume $K_n \subset U_n \subset K_{n+1}$. Then, note that if V is an open subset of K_{n+1}, the intersection $V \cap U_n$ is an open set in Ω.

Let $E \in \mathfrak{B}_\Omega$. If $\mu(E) = \infty$, only inner regularity matters, which follows from $\lim_{n\to\infty} \mu(E \cap K_n) = \infty$ and the regularity of $\mu|_{K_n}$.

Assume that $\mu(E) < \infty$ and $\varepsilon > 0$. Then, there exists n satisfying $\mu(E \backslash (E \cap K_n)) < \varepsilon/2$. Since $\mu|_{K_n}$ is regular, there exists a compact subset K of $E \cap K_n$ satisfying $\mu((E \cap K_n) \backslash K) < \varepsilon/2$, and $\mu(E \backslash K) < \varepsilon$, which shows that μ is inner regular. Next, we show outer regularity. Thanks to the regularity of $\mu|_{K_{n+1}}$, there exists an open subset V_n of K_{n+1} containing $E \cap K_n$ and satisfying $\mu(V_n \backslash (E \cap K_n)) < 2^{-n}\varepsilon$. Since $U_n \cap V_n$ is open in Ω, the union $O = \bigcup_{n=1}^\infty (U_n \cap V_n)$ is an open subset of Ω containing E, and

$$\mu(O \backslash E) \leq \sum_{n=1}^\infty \mu((U_n \cap V_n) \backslash E) \leq \sum_{n=1}^\infty \mu(V_n \backslash (E \cap K_n)) < \varepsilon.$$

Therefore, μ is outer regular.

(2) Since each K_n is separable, so is Ω, and there exists a countable dense subset $\{\omega_n\}_{n=1}^\infty$ of Ω. For each n, we choose N_n with $\omega_n \in K_{N_n}$. For $m \geq N_n$, there exists a real-valued function, $f_{n,m} \in C_c(\Omega)$, such that $f_{n,m}(\omega) = d(\omega_n, \omega)$ on K_m. The Stone–Weierstrass theorem implies that the algebra generated by $\bigcup_{n=1}^\infty \bigcup_{m=N_n}^\infty \{f_{n,m}\}$ is dense

in $C_0(\Omega)$ and, in particular, is dense in $C_c(\Omega)$. Thus, these spaces are separable.

(3) Identifying $L^p(K_n, \mu)$ with $\{f \in L^p(\Omega, \mu); \ f|_{\Omega \setminus K_n} = 0\}$, we see that $\bigcup_{n=1}^{\infty} L^p(K_n, \mu)$ is dense in $L^p(\Omega, \mu)$. As each $L^p(K_n, \mu)$ is separable, so is $L^p(\Omega, \mu)$. The fact that $C_c(\Omega)$ is dense in $L^p(\Omega, \mu)$ follows from the regularity of μ. □

If Ω is an open subset of $\mathbb{R}^n$ and μ is the Lebesgue measure, the assumption of the above theorem is satisfied. In this case, by further approximating $C_c(\Omega)$ by $C_c^{\infty}(\Omega)$, we see that $C_c^{\infty}(\Omega)$ is dense in $L^p(\Omega)$.

A.6 Stieltjes Integral

As we use the Stieltjes integral in Chapters 3 and 9, we summarize its properties here.

For a function φ defined on the finite closed interval $[a, b]$ and a partition $\Delta : a = t_0 < t_1 < t_2 \ldots t_n = b$ of $[a, b]$, we consider the quantity $\sum_{i=1}^{n} |\varphi(t_i) - \varphi(t_{i-1})|$. We denote by $V_\varphi[a, b]$ the supremum of this quantity when Δ moves over all partitions, and we call it the *total variation* of φ on $[a, b]$. When $V_\varphi[a, b]$ is finite, we say that φ is of *bounded variation*. If φ is a monotone function, we have $V_\varphi[a, b] = |\varphi(a) - \varphi(b)|$. We can see that φ is of bounded variation if and only if both the real part and the imaginary part of φ are of bounded variation.

Problem A.4. Show $V_\varphi[a, b] = V_\varphi[a, c] + V_\varphi[c, b]$ for $a < c < b$.

Let f be a bounded function on $[a, b]$, and let φ be a function on $[a, b]$ of bounded variation. We define the mesh of a partition Δ by $h(\Delta) = \max_{1 \le i \le n}(t_i - t_{i-1})$. We choose a representing point $\xi_i \in [t_{i-1}, t_i]$ from each small interval appearing in the partition Δ, and we define

$$S(f, \Delta, \{\xi_i\}, \varphi) = \sum_{i=1}^{n} f(\xi_i)(\varphi(t_i) - \varphi(t_{i-1})),$$

by replacing the length of each small interval in the Riemann sum with $\varphi(t_i) - \varphi(t_{i-1})$. If $S(f, \Delta, \{\xi_i\}, \varphi)$ tends to a certain number as $h(\Delta)$ tends to zero regardless of the choices of the partition and the representing points, we denote the number by $\int_a^b f(t)d\varphi(t)$ and call it the *Riemann–Stieltjes integral*. By the same argument as in the case of the Riemann integral, we can show that $\int_a^b f(t)d\varphi(t)$ exists for continuous f, and $|\int_a^b f(t)d\varphi(t)| \leq \|f\|_\infty V_\varphi[a,b]$ holds from the definition.

The purpose of this section is to show that we can express the Riemann–Stieltjes integral as an integral by a measure.

Lemma A.3. *For every function φ on $[a, b]$ of bounded variation, there exist monotone increasing functions φ_i, $1 \leq i \leq 4$ such that φ can be expressed as $\varphi = \varphi_1 - \varphi_2 + i(\varphi_3 - \varphi_4)$.*

Proof.   Since both the real and imaginary parts of φ are of bounded variation, it suffices to show that if φ is a real-valued bounded variation function, then it can be expressed as a difference of two monotone increasing functions. For $t \in [a, b]$, let $\psi(t) = V_\varphi[a, t]$. Then, ψ is a monotone increasing function. For $a \leq s < t \leq b$, we have

$$(\psi(t) - \varphi(t)) - (\psi(s) - \varphi(s)) = V_\varphi[s, t] - (\varphi(t) - \varphi(s)) \geq 0,$$

and $\psi - \varphi$ is monotone increasing, and $\varphi = \psi - (\psi - \varphi)$.   □

Since every monotone function is Riemann-integrable, so is every function of bounded variation. Also, recall that the product of two bounded Riemann-integrable functions are again Riemann-integrable. From these facts, we can see that the integration by parts formula

$$\int_a^b f(t)d\varphi(t) = f(b)\varphi(b) - f(a)\varphi(a) - \int_a^b \varphi(t)f'(t)dt$$

holds for $f \in C^1[a, b]$. Indeed, since

$$S(f, \Delta, \{t_i\}, \varphi) = f(b)\varphi(b) - f(a)\varphi(a) - \sum_{i=1}^n \varphi(t_{i-1})(f(t_i) - f(t_{i-1})),$$

and $\varphi(t)f'(t)$ is Riemann-integrable, the right-hand side tends to that of the integration by parts formula as $h(\Delta) \to 0$.

In what follows, we assume that φ is a monotone increasing function on $[a, b]$ and show that there exists a measure μ such that

$\int_a^b f(t)d\varphi(t) = \int_{[a,b]} f(t)d\mu(t)$ holds for every $f \in C[a,b]$. From the monotonicity of φ, the limits

$$\varphi(t-0) = \lim_{s \to t-0} \varphi(t), \quad \varphi(t+0) = \lim_{s \to t+0} \varphi(t)$$

exist and $\varphi(t-0) \le \varphi(t) \le \varphi(t+0)$ holds. If $a_1, a_2, \ldots, a_m \in [a,b]$ are discontinuous points of φ, we have

$$\sum_{i=1}^m (\varphi(a_i+0) - \varphi(a_i-0)) \le \varphi(b) - \varphi(a),$$

and the number of the points t satisfying $\varphi(t+0) - \varphi(t-0) \ge 1/n$ does not exceed $n(\varphi(b) - \varphi(a))$. Thus, the set of discontinuities of $\varphi(t)$ is at most countable, and the summation of $\varphi(t+0) - \varphi(t-0)$ does not exceed $\varphi(b) - \varphi(a)$.

We set $\varphi_r(t) = \varphi(t+0)$.

Problem A.5. Show that φ_r is monotone increasing and right continuous.

Lemma A.4. *Let φ be a monotone increasing function on $[a,b]$. Then, the following holds for every $f \in C[a,b]$:*

$$\int_a^b f(t)d\varphi(t) = f(a)(\varphi(a+0) - \varphi(a)) + \int_a^b f(t)d\varphi_r(t).$$

Proof. Since $C^1[a,b]$ is dense in $C[a,b]$, it suffices to show the equality for $f \in C^1[a,b]$. By the integration by parts formula, it suffices to show $\int_a^b (\varphi_r(t) - \varphi(t))f'(t)dt = 0$. Note that for every $\varepsilon > 0$, there exists $F \in [a,b]$ such that for every $t \in [a,b]\backslash F$, we have $|(\varphi_r(t) - \varphi(t))f'(t)| < \varepsilon$. Thus, this integral is 0. $\qquad \square$

We define $\tilde{\varphi} : \mathbb{R} \to \mathbb{R}$ by

$$\tilde{\varphi}(t) = \begin{cases} 0, & t < a, \\ \varphi_r(t) - \varphi(a), & t \in [a,b], \\ \varphi(b) - \varphi(a), & b < t. \end{cases}$$

Then, $\tilde{\varphi}$ is right continuous and monotone increasing on $\mathbb{R}$, and $\lim_{t \to -\infty} \tilde{\varphi}(t) = 0$ holds.

Lemma A.5. *Let μ be a finite Borel measure on $\mathbb{R}$ satisfying $\mu((-\infty,t]) = \tilde{\varphi}(t)$ for all $t \in \mathbb{R}$. Then, $\int_a^b f(t)d\varphi(t) = \int_{[a,b]} f(t)d\mu(t)$ holds for every $f \in C[a,b]$.*

Proof. First, note that

$$\mu(\{a\}) = \lim_{n\to\infty} \mu\left(\left(a - \frac{1}{n}, a\right]\right) = \lim_{n\to\infty}\left(\tilde{\varphi}(a) - \tilde{\varphi}\left(a - \frac{1}{n}\right)\right)$$
$$= \varphi(a+0) - \varphi(a).$$

For a partition Δ of $[a,b]$, we define a simple function f_Δ by

$$f_\Delta(t) = f(a)\chi_{\{a\}}(t) + \sum_{i=1}^{n} f(t_i)\chi_{(t_{i-1},t_i]}(t).$$

Since f is uniformly continuous on $[a,b]$, the function f_Δ uniformly tends to f as $h(\Delta) \to 0$. Thus, $\int_{[a,b]} f_\Delta(t)d\mu(t)$ tends to $\int_{[a,b]} f(t)d\mu(t)$, as μ is a finite measure. On the other hand, we have

$$\int_{[a,b]} f_\Delta(t)d\mu(t) = f(a)(\varphi(a+0) - \varphi(a)) + S(f, \Delta, \{t_i\}, \varphi_r),$$

and the statement follows from the previous lemma. $\square$

Our goal of expressing $\int_a^b f(t)d\varphi(t)$ as an integral by a measure is reduced to the following theorem.

Theorem A.10. *There exists a one-to-one correspondence between the set of finite Borel measures μ on $\mathbb{R}$ and the set of bounded right continuous monotone increasing functions $\varphi(t)$ on $\mathbb{R}$ satisfying $\lim_{t\to-\infty} \varphi(t) = 0$, and the correspondence is given by $\varphi(t) = \mu((-\infty,t])$.*

Proof. It is easy to show that the function $\varphi(t) = \mu((-\infty,t])$, for a given finite Borel measure μ, satisfies the above condition. Thus, our task is to construct μ from a given φ.

Assume that we are given a function φ satisfying the above condition. We write $\varphi(-\infty) = 0$ and $\varphi(\infty) = \lim_{t\to\infty} \varphi(t)$ as follows. Let $\mathfrak{F}$ be a finitely additive algebra of sets generated by $(-\infty,t]$, $t \in \mathbb{R}$. We denote by $\mathfrak{J}$ the set of all intervals of the form $(a,b]$

or (a, ∞), $-\infty \leq a < b < \infty$. Then, every element E of $\mathfrak{F}$ is a disjoint union of finitely many intervals $I_1, I_2, \ldots, I_n$ belonging to $\mathfrak{J}$. We would like to introduce a finitely additive set function $\mu_0 : \mathfrak{F} \to [0, \infty)$ by $\mu_0(E) = \sum_{i=1}^n \mu_0(I_i)$, $\mu_0((a, b]) = \varphi(b) - \varphi(a)$, $\mu_0((a, \infty)) = \varphi(\infty) - \varphi(a)$. We need to make sure that μ_0 does not depend on the expression of E as above, which can be done by seeing that if each I_i is further decomposed into a disjoint union of finitely many intervals belonging to $\mathfrak{J}$, the value of μ_0 remains the same. Therefore, μ_0 is well-defined. The function μ_0 is finitely additive and satisfies the following properties: $E_1 \subset E_2$ implies $\mu_0(E_1) \leq \mu_0(E_2)$, and $E \subset \bigcup_{i=1}^n E_i$ implies $\mu_0(E) \leq \sum_{i=1}^n \mu_0(E_i)$.

If μ_0 is σ-additive on $\mathfrak{F}$, then μ_0 uniquely extends to a measure μ on $\mathfrak{B}_{\mathbb{R}}$ by the Hopf extension theorem. We show this condition as follows. It suffices to show that if $E_n \in \mathfrak{F}$, $n \in \mathbb{N}$ are disjoint and $E = \bigcup_{n=1}^\infty E_n$ belongs to $\mathfrak{F}$,

$$\mu_0(E) = \sum_{n=1}^\infty \mu_0(E_n). \tag{A.1}$$

We first clarify the situation. Since we can exclude E_n with $E_n = \emptyset$ in the argument, we assume $E_n \neq \emptyset$ for every $n \in \mathbb{N}$. As $\mu_0(E_n)$ is a summation of the value of μ_0 at each connected component of E_n, to show (A.1), we may assume that each E_n is connected. If there exists n of the form $E_n = (a, \infty)$ or $E_n = (-\infty, b]$, we may show (A.1) excluding such sets from the beginning, and we may assume $E_n = (a_n, b_n]$, $-\infty < a_n < b_n < \infty$, for all $n \in \mathbb{N}$. Since each E_n is contained in a connected component of E, we can discuss each connected component E separately. Thus, we may assume that E is connected. Now, we arrive at the following situations: (1) $E = (a, b]$, $-\infty \leq a < b < \infty$, or (2) $E = (a, \infty)$, $-\infty \leq a < \infty$.

From the monotonicity of μ_0, we have $\mu_0(E) \geq \sum_{n=1}^N \mu_0(E_n)$ for every $N \in \mathbb{N}$. Note that it suffices to show

$$\forall \varepsilon > 0, \quad \mu_0(E) - \varepsilon \leq \sum_{n=1}^\infty \mu_0(E_n). \tag{A.2}$$

In the case of (2), for every $\varepsilon > 0$, there exists n satisfying $\mu_0((a, b_n]) > \mu_0(E) - \varepsilon$. If we show (A.1) only for E_k, with $b_k \leq b_n$, we get (A.2). Thus, the proof in this case is reduced to the case of (1). In the case of $a = -\infty$, a similar argument shows that the proof is reduced to the case of (1) with $-\infty < a < b < \infty$.

We assume $E = (a, b]$, $-\infty < a < b < \infty$. Since φ is right continuous, for every $\varepsilon > 0$, there exist $\delta, \delta_n > 0$ satisfying $\varphi(a+\delta) < \varphi(a) + \varepsilon/2$ and $\varphi(b_n + \delta_n) < \varphi(b_n) + 2^{-(n+1)}\varepsilon$. Then, since $[a+\delta, b] \subset \bigcup_{n=1}^{\infty}(a_n, b_n + \delta_n)$, the compactness of $[a, b]$ implies that there exist $1 \leq n_1 < n_2 < \cdots < n_N$ satisfying $[a + \delta, b] \subset \bigcup_{k=1}^{N}(a_{n_k}, b_{n_k} + \delta_{n_k})$, and $(a + \delta, b] \subset \bigcup_{k=1}^{N}(a_{n_k}, b_{n_k} + \delta_{n_k}]$ holds. Thus,

$$\mu_0(E) - \frac{\varepsilon}{2} < \mu_0((a + \delta, b]) \leq \sum_{k=1}^{N} \mu_0((a_{n_k}, b_{n_k} + \delta_{n_k}])$$

$$\leq \sum_{n=1}^{\infty} \mu_0((a_n, b_n + \delta_n]) < \sum_{n=1}^{\infty} \left(\mu_0(E_n) + \frac{\varepsilon}{2^{n+1}}\right)$$

$$= \sum_{n=1}^{\infty} \mu_0(E_n) + \frac{\varepsilon}{2},$$

and (A.2) is shown. $\qquad\qquad\qquad\qquad\qquad\qquad\qquad\square$

We call the above μ the *Lebesgue–Stieltjes measure* corresponding to φ. The integral of f by μ is sometimes written as $\int f(t)d\varphi(t)$.

A.7   Absolutely Continuous Functions

Since the class of absolutely continuous functions plays an important role in the examples treated in Chapter 10, we collect their properties here. We denote by m the Lebesgue measure of $\mathbb{R}$ in what follows.

For a continuous function f on a finite closed interval $[a, b]$ and $a \leq s < t \leq b$, we define

$$\mu_f([s, t]) = \mu_f((s, t)) = \mu_f((s, t]) = \mu_f([s, t)) = f(t) - f(s).$$

We call an open connected subset of $[a, b]$ an *open subinterval* of $[a, b]$. That is, we call any usual open interval contained in $[a, b]$, as well as $[a, s)$ and $(s, b]$, $a \leq s \leq b$, open subintervals of $[a, b]$.

Definition A.3. A function f on a finite closed interval $[a, b]$ is *absolutely continuous* if $\forall \varepsilon > 0$, $\exists \delta > 0$, whenever $I_1, I_2, \ldots, I_n$ are finitely many disjoint open subintervals of $[a, b]$ satisfying $\sum_{i=1}^{n} m(I_i) < \delta$, we have $\sum_{i=1}^{n} |\mu_f(I_i)| < \varepsilon$.

We denote by $\mathrm{AC}[a, b]$ the set of absolutely continuous functions on $[a, b]$.

Problem A.6. Show that in the above definition, even if we replace "finitely many" with "countably many", we get an equivalent definition.

The condition $f \in \mathrm{AC}[a, b]$ is equivalent to $\mathrm{Re}\, f \in \mathrm{AC}[a, b]$ and $\mathrm{Im}\, f \in \mathrm{AC}[a, b]$.

Problem A.7. Show that $\mathrm{AC}[a, b]$ is a subalgebra of $C[a, b]$.

Lemma A.6. *For* $f \in \mathrm{AC}[a, b]$, *we set* $g(t) = V_f[a, t]$. *Then,* $g \in \mathrm{AC}[a, b]$.

Proof. For any $f \in \mathrm{AC}[a, b]$ and $\varepsilon > 0$, we choose $\delta > 0$ satisfying the condition in Definition A.3. Let I_i, $i = 1, 2, \ldots, n$, be disjoint open subintervals of $[a, b]$ satisfying $\sum_{i=1}^{n} m(I_i) < \delta$. Let $\overline{I_i} = [a_i, b_i]$, and let $\Delta_i : a_i = t_{i,0} < t_{i,1} < \cdots < t_{i,m_i} = b_i$ be a partition of $\overline{I_i}$. Then, since

$$\sum_{i=1}^{n} \sum_{j=1}^{m_i} |f(t_{i,j}) - f(t_{i,j-1})| < \varepsilon,$$

we get $\sum_{i=1}^{n} V_f[a_i, b_i] \leq \varepsilon$. This means $g \in \mathrm{AC}[a, b]$. $\square$

From the above lemma, we see, as in the case of functions of bounded variation, that for every $f \in \mathrm{AC}[a, b]$, there exist monotone continuous functions $f_1, f_2, f_3, f_4 \in \mathrm{AC}[a, b]$ satisfying $f = f_1 - f_2 + i(f_3 - f_4)$.

Lemma A.7. *For a function f on $[a, b]$, the following conditions are equivalent:*

(1) $f \in \mathrm{AC}[a, b]$.
(2) *There exists $g \in L^1[a, b]$ satisfying $f(t) = f(a) + \int_a^t g(s)ds$ for all $t \in [a, b]$.*

Proof. (1) $\Longrightarrow$ (2). It suffices to show the statement assuming that f is monotone increasing. Since Theorem A.10 shows that μ_f uniquely extends to a finite Borel measure on $[a, b]$, we denote it by the same symbol μ_f. If μ_f is absolutely continuous with respect to Lebesgue measure m, the Radon–Nikodym theorem (see Exercise 2.6) implies that there exists $g \in L^1[a, b]$ satisfying $f(t) = f(a) + \int_a^t g(s)ds$. Assume that $E \in \mathfrak{B}_{[a,b]}$ satisfies $m(E) = 0$. For f and arbitrary $\varepsilon > 0$, we take δ satisfying the condition of Definition A.3. Since $m(E) = 0$, there exists an open subset U of $[a, b]$ satisfying $E \subset U$ and $m(U) < \delta$. Since U is a disjoint union of at most the countable connected components of U, and each component is an open subinterval of $[a, b]$, we get $\mu_f(E) \leq \mu_f(U) \leq \varepsilon$. As $\varepsilon > 0$ is arbitrary, we get $\mu_f(E) = 0$, and μ_f is absolutely continuous with respect to m.

(2) $\Longrightarrow$ (1). Since we can decompose g as $g = g_1 - g_2 + i(g_3 - g_4)$, $g_i \in L^1[a, b]$, $g_i \geq 0$, $1 \leq i \leq 4$, it suffices to show the statement for $g \geq 0$. Then, μ_f extends to a Borel measure on $[a, b]$, $E \mapsto \int_E g(s)ds$, and we denote it by μ_f as well. The measure μ_f is absolutely continuous with respect to m. We assume $f \notin AC[a, b]$ and deduce a contradiction in the following. If $f \notin AC[a, b]$, there exists $\varepsilon > 0$ such that for each $n \in \mathbb{N}$, there exists an open subset U_n of $[a, b]$ such that $m(U_n) < 2^{-n}$ and $\mu_f(U_n) \geq \varepsilon$ holds. Let

$$E = \bigcap_{n=1}^{\infty} \bigcup_{k=n}^{\infty} U_k.$$

Then, $m(E) = 0$ and $\mu_f(E) \geq \varepsilon$, which is a contradiction. Thus, $f \in AC[a, b]$. $\qquad \square$

For a function f on $[a, b]$ and $t \in [a, b]$, we denote by $f'(t)$ the limit

$$\lim_{h \to 0} \frac{f(t + h) - f(t)}{h}$$

if it exists. For $t = a$ or $t = b$, the limit is taken from right or left, respectively.

The fundamental theorem of calculus is generalized in the following form.

Theorem A.11. *For functions on a finite closed interval* $[a, b]$, *the following hold*:

(1) *For every* $f \in \mathrm{AC}[a, b]$, *the derivative* $f'(t)$ *exists for almost every* $t \in [a, b]$, *and* $f' \in L^1[a, b]$. *The equality* $f(t) = f(a) + \int_a^t f'(s)ds$ *holds for every* $t \in [a, b]$.

(2) *For* $g \in L^1[a, b]$ *and* $t \in [a, b]$, *let* $h(t) = \int_a^t g(s)ds$. *Then,* $h'(t)$ *exists for almost every* $t \in [a, b]$, *and* $h' = g$ *holds in* $L^1[a, b]$.

From the previous lemma, it suffices to show only (2) to prove the theorem. Since we can regard $g \in L^1[a, b]$ as an element in $g \in L^1(\mathbb{R})$ by setting $g(t) = 0$ for $t \in \mathbb{R} \backslash [a, b]$, it suffices to show the following.

Theorem A.12. *If we set* $h(t) = \int_{-\infty}^t g(s)ds$ *for* $g \in L^1(\mathbb{R})$, *the derivative* $h'(t)$ *exists for almost every* $t \in \mathbb{R}$, *and* $h' = g$ *holds in* $L^1(\mathbb{R})$.

In what follows, we prove the theorem in several steps following the argument in Rudin (1987, Chapter 7). While a different proof using a special property in dimension one is given in many books on measure theory (for example, Kolmogorov and Fomin, 1975), the following proof works in the case of $\mathbb{R}^n$ (if 3 is replaced with 3^n).

For a finite open interval $I = (a - r, a + r)$, we set $\tilde{I} = (a - 3r, a + 3r)$.

Lemma A.8. *Let* $I_1, I_2, \ldots, I_n$ *be finite open intervals. Then, there exists* $S \subset \{1, 2, \ldots, n\}$ *satisfying the following:*

- I_i, $i \in S$, *are disjoint.*
- $\bigcup_{i=1}^n I_i \subset \bigcup_{i \in S} \tilde{I}_i$.

Proof. Permuting $\{1, 2, \ldots, n\}$ if necessary, we may assume $I_i = (a_i - r_i, a_i + r_i)$ and $r_1 \geq r_2 \geq \cdots \geq r_n$. Then, note that if $i < j$ and $I_i \cap I_j \neq \emptyset$, we have $\tilde{I}_i \supset I_j$. For $1 \leq i \leq n - 1$, we set $J_i = \{i + 1, i + 2, \ldots, n\}$. We let $i_1 = 1$, and inductively let the minimal $l \in J_{i_k}$ satisfying

$$\left(\bigcup_{j=1}^k I_{i_j} \right) \cap I_l = \emptyset$$

be i_{k+1}. If there is no such $l \in J_{i_k}$, we denote k by m. Then, $S = \{i_1, i_2, \ldots, i_m\}$ satisfies the condition. $\square$

We define the *maximal function* of $g \in L^1(\mathbb{R})$ by

$$Mg(t) = \sup_{r>0} \frac{1}{2r} \int_{t-r}^{t+r} |g(s)| ds.$$

Note that $\{Mg > \lambda\}$ is an open set for $\lambda > 0$. Indeed, if $Mg(t) > \lambda$, there exists $r > 0$ satisfying

$$\frac{1}{2r} \int_{t-r}^{t+r} |g(s)| ds > \lambda,$$

and it follows from the fact that the left-hand side is a continuous function in t. Thus, Mg is lower semi-continuous and, in particular, Borel-measurable.

Lemma A.9. *We have $m\{Mg > \lambda\} \leq 3\|g\|_1/\lambda$ for all $\lambda > 0$.*

Proof. Note that if we define a measure ν on $\mathbb{R}$ by $\nu(E) = \int_E |g(s)| ds$, we get the following for $I = (t - r, t + r)$:

$$\frac{1}{2r} \int_{t-r}^{t+r} |g(s)| ds = \frac{\nu(I)}{m(I)}.$$

Let K be a compact set contained in $\{Mg > \lambda\}$. For each $t \in K$, there exists $r_t > 0$ satisfying $\nu((t - r_t, t + r_t)) > \lambda m((t - r_t, t + r_t))$. Since $K \subset \bigcup_{t \in K} (t - r_t, t + r_t)$ is an open cover of K and K is compact, there exists a finite subcover $K \subset \bigcup_{i=1}^n I_i$, $I_i = (t_i - r_{t_i}, t_i + r_{t_i})$. For this family, we choose $S \subset \{1, 2, \ldots, n\}$, as in the previous lemma, and get

$$m(K) \leq \sum_{i \in S} m(\tilde{I}_i) = 3 \sum_{i \in S} m(I_i) \leq \frac{3}{\lambda} \sum_{i \in S} \nu(I_i)$$

$$= \frac{3}{\lambda} \nu \left(\sum_{i \in S} I_i \right) \leq \frac{3\|g\|_1}{\lambda}.$$

Since K is an arbitrary compact subset of $\{Mg > \lambda\}$ and m is regular, we get $m\{Mg > \lambda\} \leq 3\|g\|_1/\lambda$. $\square$

We say that $t \in \mathbb{R}$ is a *Lebesgue point* of g if

$$\lim_{r \to +0} \frac{1}{2r} \int_{t-r}^{t+r} |g(s) - g(t)| ds = 0$$

holds.

Lemma A.10. *For $g \in L^1(\mathbb{R})$, almost every $t \in \mathbb{R}$ is a Lebesgue point of g.*

Proof. It suffices to show $m\{Tg \neq 0\} = 0$ for

$$Tg(t) = \limsup_{r \to +0} \frac{1}{2r} \int_{t-r}^{t+r} |g(s) - g(t)| ds.$$

For $n \in \mathbb{N}$, we choose $g_n \in C_c(\mathbb{R})$ satisfying $\|g - g_n\|_1 < 1/n$. Then, note that we have $Tg_n = 0$. Since

$$\frac{1}{2r} \int_{t-r}^{t+r} |g(s) - g(t)| ds \leq \frac{1}{2r} \int_{t-r}^{t+r} |g(s) - g_n(s)| ds + |g(t) - g_n(t)|$$

$$+ \frac{1}{2r} \int_{t-r}^{t+r} |g_n(s) - g_n(t)| ds,$$

we get $Tg(t) \leq M(g - g_n)(t) + |g(t) - g_n(t)|$. For $\lambda > 0$, we have

$$\{Tg > 2\lambda\} \subset \{M(g - g_n) > \lambda\} \cup \{|g - g_n| > \lambda\}.$$

Thus, letting the right-hand side be $E(\lambda, n)$, we get

$$m(E(\lambda, n)) \leq \frac{4\|g - g_n\|_1}{\lambda} \leq \frac{4}{n\lambda}.$$

Since

$$\{Tg > 2\lambda\} \subset \bigcap_{n=1}^{\infty} E(\lambda, n),$$

and the Lebesgue measure of the right-hand side is zero, we get $m\{Tg > 2\lambda\} = 0$. Since $\{Tg \neq 0\} = \bigcup_{n=1}^{\infty} \{Tg > 1/n\}$, we conclude that $m\{Tg \neq 0\} = 0$. $\square$

Proof of Theorem A.12.   Let t be a Lebesgue point of g. For $r > 0$,

$$\left| \frac{h(t+r) - h(t)}{r} - g(t) \right| + \left| \frac{h(t-r) - h(t)}{-r} - g(t) \right|$$

$$= \frac{1}{r} \left| \int_t^{t+r} (g(s) - g(t)) ds \right| + \frac{1}{r} \left| \int_{t-r}^t (g(s) - g(t)) ds \right|$$

$$\leq \frac{1}{r} \int_{t-r}^{t+r} |g(s) - g(t)| ds \to 0 \quad (r \to +0),$$

and $h'(t) = g(t)$.   □

For $f, g \in \mathrm{AC}[a, b]$, we have $fg \in \mathrm{AC}[a, b]$, and $\int_a^b (fg)'(t) dt = f(b)g(b) - f(a)g(a)$ holds. On the other hand, since we have $(fg)'(t) = f'(t)g(t) + f(t)g'(t)$ for almost every $t \in [a, b]$, we obtain the integration by parts formula:

$$\int_a^b f(t)g'(t) dt = f(b)g(b) - f(a)g(a) - \int_a^b f'(t)g(t) dt,$$

which we use in Chapter 10.

Hints for Problems and Exercises

Chapter 1

Problem 1.2 (1) From the definition of the infimum, the following holds: $\forall \varepsilon > 0$, $\exists y \in Y$, $\|[x]\| \leq \|x + y\| \leq \|[x]\| + \varepsilon$.

Exercise 1.3 The topology of Ω is given as follows:

(1) For $\omega \in \Omega$, a fundamental system of neighborhoods of ω in Ω is still the same as that in $\tilde{\Omega}$.

(2) For ∞, a fundamental system of neighborhoods in $\tilde{\Omega}$ is given by the family $\{(\Omega \backslash K) \cup \{\infty\}\}_K$, with compact $K \subset \Omega$.

Exercise 1.4 (2) μ is σ-finite. (3) For E in (2), $\chi_E \in L^p(\Omega, \mu)$.

Exercise 1.5 (1) The (compact) uniform convergence limit of a sequence of holomorphic functions is holomorphic. (2) The maximal principle.

Chapter 2

Problem 2.1 It suffices to think of a two-dimensional space.

Problem 2.2 Apply Zorn's lemma to the set of orthonormal systems in $\mathcal{H}$ with the inclusion relation.

Problem 2.3 (2) Show $\ker(I - V^*) = \{0\}$. (3) Do not forget the condition $\|a\|_2 < \infty$ when solving the equation $a = W_\alpha^* a$.

Exercise 2.2 The Cauchy–Schwarz inequality.

Exercise 2.4 (1) $f(z) = (\pi r^2)^{-1} \int_{B(z,r)} f(x+iy) dx dy$. (2) The compact uniform convergence limit of a sequence of holomorphic functions is holomorphic.

Exercise 2.5 Show $\{f_r\}_{r \in \mathbb{R}}^{\perp} = \{0\}$ by using the Fourier transform.

Exercise 2.6 (3)

$$\frac{d\nu}{d\mu} = \frac{h_\psi}{h_\varphi}.$$

Exercise 2.7 (1) Decompose Ω into countably many measurable sets with a finite measure.

Chapter 3

Problem 3.7 Take a dense countable subset $\{x_n\}_{n=1}^\infty$ of B_X, and for $\varphi, \psi \in B_{X^*}$, set $d(\varphi, \psi) = \sum_{n=1}^\infty 2^{-n} |\varphi(x_n) - \psi(x_n)|$.

Exercise 3.1 (1) The embedding $L^2(\Omega, \mu) \subset L^1(\Omega, \mu)$ is continuous (see Exercise 2.2). (2) Think of the following function:

$$f(\omega) = \begin{cases} \dfrac{\overline{g(\omega)}}{|g(\omega)|}, & |g(\omega)| > r, \\ 0, & |g(\omega)| \le r. \end{cases}$$

(3) $L^2(\Omega, \mu)$ is dense in $L^1(\Omega, \mu)$.

Exercise 3.4 (1) Since $C(\Omega)$ is separable from Theorem A.7, Problem 3.7 shows that the weak* topology on $B_{C(\Omega)^*}$ is metrizable. (2) Choose $\mu \in \mathbf{P}(\Omega)$, and set $\mu_n = n^{-1} \sum_{k=0}^{n-1} \mu \circ T^k$.

Exercise 3.5 Choose a CONS $\{e_m\}_{m=1}^\infty$ for $\overline{\text{span}\{x_n\}_{n=1}^\infty}$, and construct a subsequence $\{x_{n_k}\}_{k=1}^\infty$, by the diagonal sequence argument, such that $\{\langle x_{n_k}, e_m \rangle\}_{k=1}^\infty$ converges for every m.

Chapter 4

Problem 4.2 (1) For $\varepsilon > 0$, define a continuous function f_ε on $\mathbb{R}$ by

$$f_\varepsilon(t) = \begin{cases} -1, & t < -\varepsilon, \\ \dfrac{t}{\varepsilon}, & -\varepsilon \le t \le \varepsilon, \\ 1, & \varepsilon < t, \end{cases}$$

and consider $f_\varepsilon(D_N(t))$.
(2)

$$\|D_N\|_1 = \frac{1}{\pi} \int_0^\pi \left| \frac{\sin(N + \frac{1}{2})t}{\sin \frac{t}{2}} \right| dt \ge \frac{2}{\pi} \int_\pi^{N\pi} \frac{|\sin s|}{s} ds.$$

Exercise 4.2 The closed graph theorem.

Exercise 4.3 (1) The Banach–Steinhaus theorem. (3) Show that

$$\left\{ \frac{1}{h_n^2} \left(f(\zeta + h_n) - f(\zeta) - h_n g(\zeta) \right) \right\}_{n=1}^{\infty}$$

is bounded by using the principle of uniform boundedness.

Exercise 4.4 If $x, y \in \ell^2(\mathbb{Z})$ satisfy $x_k = y_k = 0$ except for finitely many k, we have $\langle U^n x, y \rangle = 0$ for sufficiently large n.

Exercise 4.5 (2) $\mathcal{R}(I - U)^{\perp} = \ker(I - U^*) = \ker(I - U)$.

Chapter 5

Problem 5.2 Let $x, y \in \overline{C}$, and let $0 < t < 1$. For every neighborhood U of $tx + (1 - t)y$, there exist a neighborhood V of x and a neighborhood W of y satisfying $tV + (1 - t)W \subset U$.

Problem 5.4 The condition for equality of the Minkowski inequality.

Exercise 5.2 (1),(2) $C_c(\mathbb{R})$ is dense in $L^2(\mathbb{R})$.

Exercise 5.3 (2) We can define the support of a regular Borel measure (see Section A.5).

Exercise 5.4 (2) $\implies$ (1). Let $\mu = t\mu_0 + (1 - t)\mu_1$, $0 < t < 1$, $\mu_0, \mu_1 \in \mathbf{P}_T(\Omega)$. Then, μ_i is absolutely continuous with respect to μ, and the T-invariance implies

$$\mu \left\{ \frac{d\mu_i}{d\mu} \circ T \neq \frac{d\mu_i}{d\mu} \right\} = 0.$$

Chapter 6

Exercise 6.1 See Problem 2.4. Compute it by using $L^2(\mathbb{T})$.

Exercise 6.2 $\mathcal{H} = \mathbb{C}^2$ is enough.

Exercise 6.6 The function $f(z) = z$ is not invertible as an element in $A(\mathbb{D})$, but it is invertible as an element in $C(\partial\mathbb{D})$.

Chapter 7

Problem 7.2 (1) Take a basis $\{x_i\}_{i=1}^n$ for M and its dual basis $\{x_i^*\}_{i=1}^n \subset M^*$ (that is, $x_j^*(x_i) = \delta_{i,j}$). Extend x_i^* to $\varphi_i \in X^*$ by the Hahn–Banach extension theorem, and set $N = \bigcap_{i=1}^n \ker \varphi_i$.

Exercise 7.1 To show that C is totally bounded, first for a given $\varepsilon > 0$, choose N satisfying $\sum_{n=N+1}^{\infty} n^{-2} < \varepsilon^2/2$.

Exercise 7.2 (2) To show that $TB_{L^2[0,\infty)}$ is totally bounded, for a given $\varepsilon > 0$, apply the Ascoli–Arzelà theorem to the restriction of $TB_{L^2[0,\infty)}$ to $[0, r]$ for sufficiently large $r > 0$.

Exercise 7.3 (1) First, show it for a trigonometric polynomial f. Note that the map $C(\mathbb{T}) \to \mathbf{B}(L^2(\mathbb{T}))$, $f \mapsto [M_f, P_+]$, is continuous. (2) Through the identification of $\mathbf{B}(H^2(\mathbb{T}))$ and $P_+\mathbf{B}(L^2(\mathbb{T}))P_+$, the Toeplitz operator T_f is identified with $P_+M_fP_+$. Then, $T_{fg} - T_fT_g = P_+M_f(I - P_+)M_gP_+$. (3) Use the Atkinson theorem.

Exercise 7.4 (2) $\|T_f h_r\|_2 \le \|M_f h_r\|_2$. (3) If $T_f \in \mathrm{FR}(H^2(\mathbb{T}))$, the Atkinson theorem implies that there would exist $S \in \mathbf{B}(H^2(\mathbb{T}))$ and $K \in \mathbf{K}(H^2(\mathbb{T}))$ satisfying $ST_f = I + K$.

Exercise 7.5 (3) See Exercise 2.1. (4) The continuity of the index.

Exercise 7.6 (1) Since $\lambda I - V$ is unitarily equivalent to $T_{\lambda - e_1}$, it suffices to see whether $T_{\lambda - e_1} \in \mathrm{FR}(\ell^2)$ or not. (2) Show $\mathrm{ind}(\lambda I - V) = -1$ for $\lambda \in \mathbb{C}$ with $|\lambda| < 1$.

Chapter 8

Problem 8.1 (1) The right-hand side is $\min_{\dim \mathcal{K} = n-1} \|T - TP_{\mathcal{K}}\|$.

Problem 8.3 Compute the integral kernel of VV^*.

Exercise 8.3 Compute the maximal eigenvalue of B_θ.

Exercise 8.4 (1) Compute $A_k e_n$.

Exercise 8.5 (1) See the proof of Proposition 8.1. (2) Refer to the proof of the well-known fact that the Fourier series of $f \in C^1(\mathbb{T})$ absolutely and uniformly converges.

Exercise 8.6 (1) $\|[M_f, P_+]\|_{\mathrm{HS}}^2 = \sum_{m,n \in \mathbb{Z}} |\langle [M_f, P_+]e_m, e_n \rangle|^2$. (2) $T_fT_g - T_{fg} = [M_f, P_+](I - P_+)[M_g, P_+]$. (3) First, show it for trigonometric polynomials f and g. For the general case, use the estimate in (2).

Exercise 8.7 (1) See the proof of Lemma 7.5.

Chapter 9

Problem 9.1 (1) Since $\ker A^\perp = \overline{\mathcal{R}(A)}$, it suffices to show that $\{\|A^{1+\frac{1}{n}} - A\|\}_{n=1}^\infty$ converges to 0. (2) Let B be the strong limit. Then, it is a projection satisfying $AB = B$.

Problem 9.2 First, show that the function $\omega \mapsto d(\omega, F)$ is continuous. Think of $e^{-nd(\omega, F)}$.

Exercise 9.2 (1) The set of polynomial functions is dense in $C(\sigma(A))$. (2) The binomial coefficients. (3) The Catalan number.

Exercise 9.3 (1) Let $A = \int_{\sigma(A)} t\, dE_t^A$ be the spectral decomposition. For every $\varepsilon > 0$, we have $\dim \mathcal{R}(E_{\lambda+\varepsilon}^A - E_{\lambda-\varepsilon}^A) = \infty$. (2) The sequence $\{e_n\}_{n=1}^\infty$ in (1) weakly converges to 0. (3) Show $\lambda I - A \in \mathrm{FR}(\mathcal{H})$.

Exercise 9.4 (1) Let $A = \int_{\sigma(A)} t dE_t^A$ be the spectral decomposition of A. For polynomials f and g, we have

$$\langle f(A)x_0, g(A)x_0 \rangle = \int_{\sigma(A)} f(t)\overline{g(t)}d\langle E_t^A x_0, x_0 \rangle.$$

(2) By using Zorn's lemma, show the existence of a subset $\{x_i\}_{i \in I}$ of $\mathcal{H}$ such that $\mathcal{H}_{A,x_i}$ are mutually orthogonal, and span($\bigcup_{i \in I} \mathcal{H}_{A,x_i}$) is dense in $\mathcal{H}$.

Exercise 9.5 (1) The Cauchy–Schwarz inequality holds without the condition "$\|x\| = 0 \iff x = 0$".

Chapter 10

Problem 10.1 (1) $D^{**} \subset D_c \subset D^*$ implies $D^{**} \subset D_c^* \subset D^*$.
(2) Since the linear map $\Phi : \mathcal{D}(D^*) \to \mathbb{C}^2$, $\Phi(f) = (f(0), f(1)) \in \mathbb{C}^2$, is a surjection with $\ker \Phi = \mathcal{D}(D^{**})$, there exists a one-to-one correspondence between the set of operators T satisfying $D^{**} \subsetneq T \subsetneq D^*$ and the set of one-dimensional subspaces in $\mathbb{C}^2$.

Problem 10.2 Solve the differential equation $f'' + \lambda f = 0$, with the boundary condition corresponding to T^*T.

Problem 10.3

$$\left\| \frac{1}{it}(e^{itA}x - x) - Ax \right\|^2 = \int_{\mathbb{R}} \left| \frac{e^{it\lambda} - 1}{it} - \lambda \right|^2 d\langle E_\lambda^A x, x \rangle$$

$$= \int_{\mathbb{R}} \left| \int_0^1 (e^{ist\lambda} - 1)ds \right|^2 \lambda^2 d\langle E_\lambda^A x, x \rangle.$$

Exercise 10.1 (1) For $\lambda \in \mathbb{C} \backslash \sigma_p(D_c)$, solve the differential equation $if'(t) + \lambda f(t) = g(t)$, with the boundary condition $f(1) = cf(0)$ ($f(0) = 0$ for $c = \infty$), and express $(\lambda I - D_c)^{-1}$ as an integral operator.

Exercise 10.2 (1) For $g \in \mathcal{D}(L^*)$ and $L^*g = g^*$, let $h(t) = \int_0^t (t - s)g^*(s)ds$. Then, $-\langle f'', g \rangle = \langle f, h'' \rangle = \langle f'', h \rangle$ holds for every $f \in C_c^\infty(0, 1)$. (5) For every $\lambda \in \mathbb{C} \backslash \sigma_p(L')$, solve the differential equation $\lambda f + if'' = g$, with the boundary condition corresponding to L', and express $(\lambda I - L')^{-1}$ as an integral operator.

Exercise 10.4 (2) $\eta = 0$. (3) $\zeta \in \mathbb{R}$ and $\xi\eta = \zeta(\zeta + 1)$.

Exercise 10.5 (1) $\mathcal{R}(\tilde{D}^{**} - zI)^\perp = \ker(\tilde{D}^* - \bar{z}I)$. (2) Solve the differential equation $zf + if' = g$.

Exercise 10.6 (2) We have $T^*y = T^*(I + TT^*)(I + TT^*)^{-1}y$ for $y \in \mathcal{D}(T^*)$.

Exercise 10.7 First, compute

$$\lim_{\varepsilon \to +0} \frac{1}{\pi} \int_{-\infty}^{t} \frac{\varepsilon}{(s - \lambda)^2 + \varepsilon^2} ds.$$

Appendix A

Problem A.3 The case of $p = \infty$ is easy. For $p \neq \infty$, let $s = p/r$, $b_i = |a_i|^r$, and show $\|b\|_s \leq \|b\|_1 \leq n^{1-1/s}\|b\|_s$.

Bibliography

Akhiezer, N. I. and Glazman, I. M. (1981). *Theory of Linear Operators in Hilbert Space*. Vol. I, II. Translated from the third Russian edition by E. R. Dawson. Translation edited by W. N. Everitt. Monographs and Studies in Mathematics. Pitman (Advanced Publishing Program), Boston, Mass., London, pp. 9–10.

Carothers, N. L. (2005). *A Short Course on Banach Space Theory*. London Mathematical Society Student Texts. Cambridge University Press, Cambridge, p. 64.

Conway, J. B. (1990). *A Course in Functional Analysis*, Second edition. Graduate Texts in Mathematics. Springer-Verlag, New York, p. 96.

Dunford, N. and Schwartz, J. T. (1988). *Linear Operators. Part I. General Theory*, with the assistance of W. G. Bade and R. G. Bartle. Reprint of the 1958 original. Wiley Classics Lib. Wiley-Intersci. Published by John Wiley & Sons, Inc., New York, 1988.

Folland, G. B. (1999). *Real Analysis*, Second edition. Modern Techniques and their Applications. Pure Appl. Math. (N.Y.). Wiley-Intersci. Published by John Wiley & Sons, Inc., New York, 1999.

Gohberg, I. C. and Krein, M. G. (1969). *Introduction to the Theory of Linear Nonselfadjoint Operators*. Translated from the Russian by A. Feinstein. Transl. Math. Monogr., Vol. 18. American Mathematical Society, Providence, RI, 1969.

Jordan, P. and von Neumann, J. (1935). On inner products in linear, metric spaces, *Annals of Mathematics*, **36**(3), 719–723.

Kadison, R. V. and Ringrose, J. R. (1997). *Fundamentals of the Theory of Operator Algebras*, (Vol. I): Elementary theory. Reprint of the 1983 original. Graduate Studies in Mathematics. American Mathematical Society, Providence, RI, p. 15.

Kelley, J. L. (1975). *General Topology*. Reprint of the 1955 edition (Van Nostrand, Toronto, Ont.). Graduate Texts in Mathematics, No. 27, Springer-Verlag, New York-Berlin.

Kolmogorov, A. N. and Fomin, S. V. (1975). *Introductory Real Analysis.* Translated from the second Russian edition and edited by R. A. Silverman. Corrected reprinting. Dover Publications, Inc., New York.

Pedersen, G. K. (1989). *Analysis Now.* Graduate Texts in Mathematics. Springer-Verlag, New York, p. 118.

Rudin, W. (1986). *Real and Complex Analysis,* Third edition. McGraw-Hill Book Co., New York.

Schmüdgen, K. (2012). *Unbounded Self-adjoint Operators on Hilbert Space.* Graduate Texts in Mathematics. Springer, Dordrecht, p. 265.

Simon, B. (2005). *Trace Ideals and Their Applications,* Second edition. Mathematical Surveys and Monographs. American Mathematical Society, Providence, RI, p. 120.

Index

www.ingramcontent.com/pod-product-compliance
Lightning Source LLC
Chambersburg PA
CBHW070214190526
45161CB00002B/80